LINKAGE PERSPECTIVE IN AGRICULTURAL EXTENSION

Shantanu Kumar Dubey
Uma Sah
A.K. Singh

2018

Daya Publishing House®

A Division of

Astral International Pvt. Ltd.

New Delhi - 110 002

First Published, 2011
Reprinted, 2018

ISBN: 978-81-7035-762-9 (Hardbound)

Publisher's note:

Every possible effort has been made to ensure that the information contained in this book is accurate at the time of going to press, and the publisher and author cannot accept responsibility for any errors or omissions, however caused. No responsibility for loss or damage occasioned to any person acting, or refraining from action, as a result of the material in this publication can be accepted by the editor, the publisher or the author. The Publisher is not associated with any product or vendor mentioned in the book. The contents of this work are intended to further general scientific research, understanding and discussion only. Readers should consult with a specialist where appropriate.

Every effort has been made to trace the owners of copyright material used in this book, if any. The author and the publisher will be grateful for any omission brought to their notice for acknowledgement in the future editions of the book.

Published by : **Daya Publishing House®**
A Division of
Astral International Pvt. Ltd.
– ISO 9001:2015 Certified Company –
4736/23, Ansari Road, Darya Ganj
New Delhi-110 002
Ph. 011-43549197, 23278134
E-mail: info@astralint.com
Website: www.astralint.com

PREFACE

Development of agriculture and livestock sectors in India has direct implications on country's overall economic growth. Application of agricultural technologies by the end users is standard yardstick to quantify agricultural development in a given time and space. However, timely application of agricultural production technology has the critical antecedents like strong extension-farmer linkage besides other factors like input availability, marketing infrastructure, transport facility, etc. Further, the design, testing, validation and delivery of appropriate technology demand mutual and reciprocal interaction between research, extension and farmers. Such demand becomes more pronounced in case of dairy development which is more technology sensitive and input demanding.

Hence, the book at hand attempts to highlight the all possible dimensions of linkage perspective in agriculture. This include the conceptual framework of linkages, linkage typology and their operationalization, need to evolve organizational linkage mechanism, linkage constraints, factors affecting linkages, etc. Besides, the book also contains the methodology to develop the measuring instrument to quantify the strength of research-extension-farmer linkage. The available knowledge base on various conceptual and empirical dimensions of linkage related issues has been reviewed and synthesized in the meaningful manner in the book. The detailed empirical investigation on linkages among research, extension and dairy farmers operating under various organization systems have been narrated to consolidate the case for institutionalization of adequate structural linkage mechanism with particular references to dairy development. Therefore, the effectiveness of prevailing structural linkage mechanism has also been documented and analyzed. The different socio-personal, psychological, organizational, external and constraining factors are analyzed objectively and statistically. Based on the findings, an empirical model depicting the linkage strength between all three partners and the significant variables predicting linkage strength has been evolved and guideline for effective linkage mechanism has also been suggested.

Authors are confident that this book will be extremely useful for the extension students, researchers, development practitioners as well as managers of agricultural development in general and dairy development in particular.

Shantanu Kumar Dubey
Uma Sah
A.K. Singh

CONTENTS

ABBREVIATIONS

AEOFL	=	Average Extent of Overall Functional Linkage
AKIS	=	Agricultural Knowledge and Information System
ANOVA	=	Analysis of Variance
BAU	=	Birsa Agricultural University
CCS HAU	=	Chaudhary Charan Singh Haryana Agricultural University
CV	=	Coefficient of Variation
DF	=	Dairy Farmers
DKIS	=	Dairy Knowledge and Information System
DOE	=	Directorate of Extension
ECL	=	Extent of Communication Linkage
EFLT	=	Extent of Functional Linkage in Training
ELIE	=	Extent of Linkage in Implementation and Evaluation
ELPDM	=	Extent of Linkage in Planning and Decision Making
ELSS	=	Extent of Linkage in Study and Services
EP	=	Extension Personnel
FSRP	=	Farming Systems Research Project
GBPUA&T	=	Govind Ballabh Pant University of Agriculture & Technology
IARI	=	Indian Agricultural Research Institute
ICDP	=	Intensive Cattle Development Project
IIHR	=	Indian Institute of Horticultural Research

IIM	=	Indian Institute of Management
IVLP	=	Institute Village Linkage Programme
IVRI	=	Indian Veterinary Research Institute
JIE	=	Joint Implementation and Evaluation
JPDM	=	Joint Planning and Decision Making
KGK	=	Krishi Gyan Kendra
KVK	=	Krishi Vigyan Kendra
MEFL	=	Mean Extent of Functional Linkage
NAEP	=	National Agricultural Extension Project
NARP	=	National Agricultural Research Project
NATP	=	National Agricultural Technology Project
NDRI	=	National Dairy Research Institute
OEFL	=	Overall Extent of Functional Linkage
ORP	=	Operational Research Project
RAJCOVAS	=	Rajiv Gandhi College of Veterinary and Animal Sciences
RDFP	=	Recommended Dairy Farming Practices
RE	=	Research Extension
RP	=	Research Personnel
SD	=	Standard Deviation
SDAH	=	State Department of Animal Husbandry
SDFP	=	Scientific Dairy Farming Practices
SLM	=	Structural Linkage Mechanism
UAS	=	University of Agricultural Sciences

1 Systems Perspective in Research, Extension and Farmers Interaction

Systematic and empirical analysis of the complex interrelationship and linkage pattern of research, extension and clients call for sound comprehension of its theoretical base. The behavioural scientists and communication scholars have tried to elucidate this linkage pattern with respect to systems perspective of communication behaviour of research, extension and client. The chapter has been organized under the following sub heads:

1.1 Systems Involved in Development

System induced in development hierarchy through which ideas, new knowledge and farm information find expression in the practical needs of the clients have been conceptualized by different authors differently. Rogers and Yost (1960) while studying the communication behaviour of county extension agents proposed a two step flow of communication model from scientist to farmers and recognized three elements namely, (a) Agricultural scientists, (b) County extension agents, and (c) Farm people.

Coughenour (1968) described three systems with reference to diffusion of innovation and social action. These are : (a) Innovation developing system, (b) Innovation disseminating system, and (c) Innovation using system. Guba (1968) conceptualized three stages in the knowledge development and diffusion, namely, (a) Knowledge production, (b) Knowledge of dissemination, and (c) Knowledge utilization.

Lionberger and Chang (1970) identified three distinct special systems in their study of farm information for modernizing agriculture in Taiwan. These are : (a) Innovating system, consisting of agricultural scientists, (b) Disseminating system, consisting of communicators and change agents, and (c) Users comprising of farm operation. Jain (1970) proposed a theoretical model of research dissemination and utilization process consisting of three social systems. These are : (a) Research system, responsible for developing research knowledge, (b) Linker system performing the role of linking research and client system, and (c) Client system, which adopts and integrate the research knowledge.

Axinn (1972) in the study of modernizing world agriculture, a comparative study of extension education system, described three systems and these are : (a) Research system, (b) Extension education or change system, and (c) The target system. Further, Singh and Kumar (1973) conceptualized these three systems as suggested by Guba (1968). Akhouri (1973), while studying the communication behaviour of extension personnel conceptualized the following three systems, namely (a) Research system, (b) Agricultural extension education system, and (c) Client system. Ambastha (1974) conceptualized these systems as the innovation development system, extension system and the client system.

From the above synthesized information, it is aptly observed that almost all the authors have viewed the systems of development hierarchy with mainly three elements. These three elements have been named in one or other way by the different scholars, but connote the similar meaning. Even at the present point of time neither of these elements has been excluded. However, understanding the mode and intensity of interaction among these elements have witnessed a tremendous evolution. Some of these trends have been explained in Chapter 2.1. It could be argued that development being dynamic in nature, the variable for the interaction among the elements of development hierarchy must change with the change in time.

Having known the elements/systems of development, it is required to further elaborate these elements separately. In the following, sub heads these systems have been described.

(*a*) Research System

This system is responsible for the creation of new information from which new technology emerged. Such creation of knowledge was often supposed to response to farmers' important felt needs and felt problems. But, the research system work somewhat in isolation as observed by Dwarkinath and Channegowda (1974), Clausen (1984) and Cernea *et al.* (1985). The reasons for such isolation were limited understanding of extension needs and practices by the research personnel and inadequate feedback of farm production problem to the research as understood by these authors.

In the present book, research system has been used as more useful nomenclature for understanding and describing the system in development which is engaged in more of applied researches in order to generate useful information and knowledge in the field of animal production and veterinary. Two types of research systems namely, NDRI and HAU which are operating in the State of Haryana have been included for empirical study.

(*b*) Extension System

This system operates between the research system and client system. The members of the extension system are change agent, input suppliers, bankers, etc. These members are entrusted to process the agricultural information received from research system and their communication and promotion to the users. Apart from that, they also receive necessary feedback of farmers needs, problem and perception of the technology/ practices. These feedbacks are either entertained by the members of this system or passed on to the research system for further processing.

Dwarkinath and Channegowda (1974) observed that the extension system was not entirely composed of persons who were equipped to do this job efficiently. Their knowledge and skills of both the technology and manner of its transmission needed improvement.

In the similar line, a number of authors have revealed non-availability of well trained personnel to provide the intensity of services needed (Mamoria, 1966; Lerner and Schramn, 1967; Byrnes, 1968), ill equipped extension machinery (Baweja, 1974) and hence, non-aversion of much of resistance on the part of farmers (Byrnes, 1968).

Present book attempts to analyze three types of extension systems, *viz.*, NDRI, HAU and SDAH for investigating their strength of linkages with selected research systems and the client systems under their areas of operation. The term 'extension personnel' was used for all those who were engaged in extension education activities and technology transfer. They were performing jobs either an extension administrator or as a field functionary.

(*c*) Client System

This system consisted of clientele groups concerned with the utilization of technology. The Joint Indo American Study Team (1970) laid down that the impact of extension education programme on rural countryside depended not only upon the skills of extension workers and innovativeness of the University staff, but also upon the responsiveness of the rural population and the strength of the supporting agencies. The socio economic factors, cultural set up and infrastructural scenario also constitute important parameters of client system. Client system in this book is the beneficiary dairy farmers of the selected extension systems.

1.2 Concept of Linkages.

An examination of the above discussed research system, extension system and client system suggests that these systems are dependent on each other. Not only this, every basic component of these systems need to interact and enrich one another in a way that every action adds to the strength of the other system. The probable interactional mechanism between these three systems in terms of their members, functions and inter-relationship in the process of farm technology development, dissemination and utilization as advocated by Singh (1975) could be depicted below:

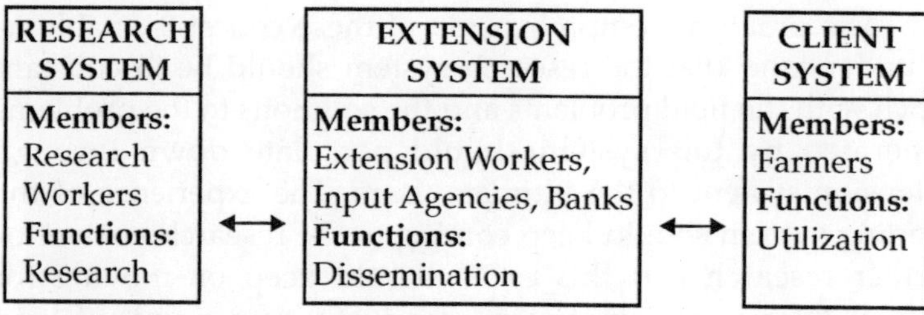

Theoretically, such model may look easy to be accomplished, but the experience has shown that it is quite difficult to achieve on floor.

Venkatesan (1985) while discussing on improving research extension linkage observed that a range of problem that surface with an effective extension system in place and that require strong research-extension linkage, which according to the author was acknowledged as crucial to the success of any extension system. The author opined that there were many factors that influence the extent of this linkage. Unless these factors were analyzed, it was not possible to establish firm and enduring interaction. The author further pleaded that linkage should not depend on the chance contact, at the personal level, between research scientist and the extension worker deployed at a place. Rather, linkage should be an integral part of the system.

Linkage in the transfer of technology, in reality, constitute a type of circulatory system, which to enthuse life and health is a body, has to work round the clock on sustained basis with utmost efficiency.

Linkages has been conceptualized by Axinn (1988) as the set of channels which act as the means by which people in organization communicate. This organizations could be research and extension organization. Prasad (1988) viewed linkage as a two way communication and interaction process among the three actors of development, *viz.*, research, extension and farmers. He further argued that linkage embraces the components of organizational variables and individual behavioural variables.

The practical comprehension of these concepts enable us to understand that the research system should be in constant touch with the field problems and the solutions to the problems generated by this system should percolate down through extension system to the ultimate users. The experience of the farming system should keep coming to the research station for further research and this circle should keep on moving for mutual interest and increasing productivity per unit of land and animal. Weakness in linkage in any of systems will throttle the research system and with the passage of time, the fear of becoming research system simply a ritual can not be ruled out. Obviously, under such a state of affairs, the linkage become not only weak, may also get lost.

As a further discussion, Singh *et al.* (1991) conceptualized linkage to have both institutional and functional relationships. Institutional linkage refers to any interaction which may exist between research institutions/departments, and extension agencies. The functional linkage focus on particular function that need to be performed by the system to link technology generation with technology transfer activity.

In another instance, Prasad and Reddy (1991) viewed linkages as a mental function. Individual must appreciate research and development alike linkage is a social process and its subjective nature make it all the more difficult to visualize elements of integration process, use and manipulate them, singly or in combination, in actual agricultural situations. Its complexity is further pronounced because of varied backgrounds of the persons involved in the process. The biological and physical scientists usually do not understand and appreciate the sociological processes required to integrate their efforts and programmes (Anonymous, 1992).

From the above arguments, it can be safely arrived at, that linkages at every level in the process of technology transfer provide the structure and framework for research system, extension system and clientele system. Realizing this importance, an in built mechanism should be ingrained in the organization responsible for technology transfer not only for checking the weakness at every interactional level, but also for offering remedial measures required for strengthening the linkages.

1.3 Linkages Typology

An empirical probe of any stimuli, object or event warrants a clear cut understanding of that subject. In this context, hence, linkages can not be the exception. In the present sub head, therefore, attempt has been made to understand the various types of linkages as explained by few authors.

Esman and Blaise (1966) had identified following four types of linkages for any extension system:

(*a*) **Enabling linkage** : It means the linkage of extension system with government machinery, cooperatives and private sectors.

(*b*) **Functional linkage** : It is the linkage of extension system with the research system.

(*c*) **Normative linkage** : It is the linkage of extension system with colleagues in other related profession.

(*d*) **Diffuse linkage** : It means the linkage of extension system with the farmers or clients.

These four types of linkages for extension system are not supposed to be mutually exclusive. These are, rather, complementary and supplementary to each other as well as to the extension system.

Another group of authors named Prasad (1988) and Singh (1994) had made following typology of the linkage:

(*a*) **Structural linkage** : It refers to the provision of formal mechanism which may exist between research institutions, departments, and extension agencies to promote interaction among them. These resources, responsibility for collaboration and other forum that has been created for them to coordinate their activities.

(*b*) **Functional linkage :** It focus on a particular function that need to be performed by the system/organization/department to link technology generation, technology transfer and clients. This include number of sub parameters upon which functional linkage is ascertained.

In the present book, the typology as suggested by Singh (1994) has been used. The details about the parameters of functional linkage have been elaborated in methodology adopted as mentioned in Chapter 6.

1.4 Need to Evolve Linkage

No any organization/institution/department can survive and work in isolation. It has some accepted or non accepted linkages which are vital for its survival. However, the nature, dimension and mechanism of linkage would differ according to the need and objectives of the institute/department, but at certain points, areas of operation and responsibilities they have some common goal and objectives to achieve. Hence, by exploring the linkage on common goal, objectives and activities, the conflict and overlapping of work can be reduced and effective relationship can be maintained.

The need for effective two way linkages among agricultural and livestock research, extension and clients is beyond any dispute. Although, researchers may have contact with farmers in order to directly acquaint with laters' production conditions and technology requirements, former scarcely have time for extension contacts with farmers. Similarly, extension would have little to offer in the long run without research input from research. Notwithstanding this mutual dependence of research and extension, in many places the linkage between them are weak.

Whyte (1975) stated that unless these organizations (research and extension) are effectively linked together, little of their output ever gets to the farmers. Pickering (1985) reported the requirements of broad consensus with Asian research and extension. Claar and Bentz (1984) noted the need for a special linkage between research and extension, because most of the information that extension transmits to clients originate outside the extension organization.

The World Bank (1985) identified the gap between research and extension as the most serious institutional problem to be overcome in developing an effective R&E system. Uphoff *et al.* (1983), Venketasan (1985) and Johnsan and Claar (1986) also held the same opinion and noted the need for a strong research and extension linkage.

All over the world, in most of the projects and programmes, there are provisions for linking research and development efforts for multiplying gains in terms of coordination through committees, planning forums, evaluation team, diagnostic groups, etc. However, in operational aspects of linkage has become a continuing problems and a current topic for discussion everywhere.

Linkage is a multiple communication process. In a field like livestock and dairying, there are inter disciplinary and multi institutional requirements. For a proper functioning of linkage, understanding of linkage processes and elements of linkage is essential.

Linkage as a connecting channel between two major units/ organizations has been dealt in most of the papers presented in 17th convention of Agricultural Universities. The recommendations also spelled out the types of linkages of the respective SAU with the development agencies at the state and central levels.

The achievements of SAUs and ICAR institutes in generating crop and livestock production technologies can be attributed to their effective linkages with the SDA and SDAH. In recent years, the importance of establishing new linkages has been realised in terms of area development activities. The linkages in different forms among the relevant components of subsystems, enable to coordinate the efforts, get mutual support, avoid duplication and overlapping of activities in pursuit of common goals.

Some of the relationship have been clearly indicated in the recommendations under statutory provision and under organizational set up. It is further recommended that not only the provisions be materialized, also a beginning on as many of these as possible should be made straight way. As far as possible, Memoranda of Understanding (MoU) should be made between the University/Institute and the concerned department should be evolved and put to work.

1.5 Conceptual Framework

The empirical investigation of linkage in this book has been conceptualized three dimensional. In the first instance, efforts were putforth to understand and study the structural mechanism existing in the selected organizational systems fostering the linkage of different types, *i.e.*, Research-Extension, Extension-Farmers and Research-Farmers. The structural linkages were explored in terms of various arrangements, *i.e.*, Research council, Extension council, Research advisory committee, meetings, camps, adaptive research, trial, etc.

Further, the strength of reciprocal functional linkage between research and extension, extension and farmers, and research and farmers as perceived by each element was studied. The strength of research and extension linkage was studied in various organizational arrangements of research and extension. In general, two types of such arrangements were noted. Firstly, when both research and extension were operating from same organization. Under this concept, research personnel from the departments of animal production like Division of Dairy Cattle Breeding, Dairy Cattle Nutrition and Animal Physiology (from NDRI) and from the College of Animal Sciences and College of Veterinary Science (from HAU) were taken. The extension personnel were selected from Directorate of Extension to field level (from HAU) and from NDRI, TOT wings in terms of KVK, ORP/FSR, IVLP and Division of Dairy Extension were selected in order to sample the required number of extension personnel.

The second arrangement/system of research and extension was when both of them were operating from separate organization. Hence, the State Department of Animal Husbandry (SDAH) was included in the study to meet the requirements. From SDAH, extension personnel were taken to assess their extent of functional linkage with research personnel operating in NDRI and HAU, and vice-versa. The details of the parameters of functional linkage between research and extension are described in methodology chapter.

The strength of functional linkage between extension and farmers were also studied. Since, the study included three types of extension systems, *viz.*, NDRI, HAU and SDAH, the dairy farmers were sampled from these three systems. A comparative analysis of extent of linkage between extension and farmers was carried out for more clarity and better understanding. The indicators/parameters used for measuring extension-farmer linkage are elaborated in methodology chapter. Similarly, the functional linkage (in terms of communication linkage only) between research and farmers was also studied. The second dimension of the present study was to identify the factors affecting the strength of various types of linkages. A number of variables, in terms of personal, psychological and organizational variables were taken to identify the factors affecting the linkage between research and extension. These variables were studied subjectively as well as statistical treatments were done to identify the factors objectively. From the farmers' point of view, socio personal, socio economic, psychological, communication and some other relevant variables were taken to identify the factors affecting their extent of linkage with extension personnel. These variables are described in the methodology chapter. There is not a single linkage related activities, which is operating in the absence of any bottleneck. In the present study, hence, attempt was made to delineate the constraints of various types impeding the different activities of linkages. The constraints were identified as perceived by the three actors of development, *viz.*, research personnel, extension personnel and farmers. The above discussed conceptual framework of the present study is diagrammatically depicted in Figure 1.

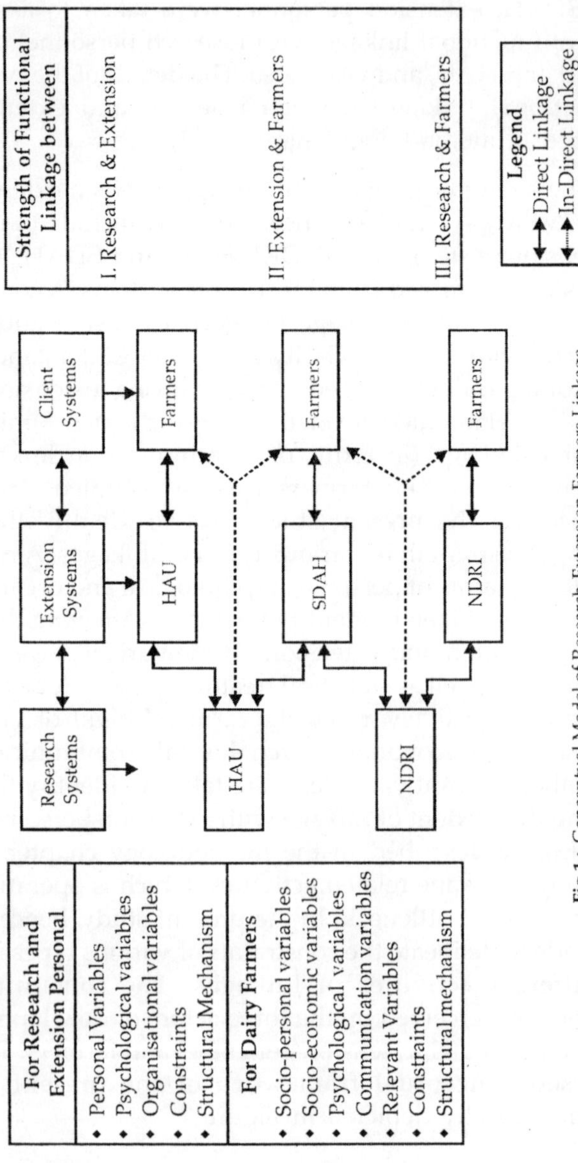

Fig.1 : Conceptual Model of Research-Extension-Farmers Linkage

2 Available Knowledge Based on Research-Extension-Farmers Linkages

The review of available conceptual and empirical information helps to consolidate the theoretical comprehension of any subject and also guides to identify the gaps in the areas of past researches. Though, there were scanty available empirical evidence which could be directly associated with the present context, albeit, attempts were endeavoured to synthesize some of the literature pertinent to present chapter. Relevant literature have been reviewed and presented under the following sub heads:

2.1 Status and trends of communication research

2.2 Linkages between research, extension and farmers

  2.2.1 Linkages between research and extension

  2.2.2 Linkages between extension and farmers

  2.2.3 Linkages between research and farmers

2.3 Variables/parameters for measuring linkages between research, extension and farmers

2.1 Status and Trends of Communication Research

Since sixties, a large number of studies have been undertaken on farmers, extension and research interaction, but the system approach was missing and the studies were on communication behaviour of different segments of farmers extension-research continuum. Most of the Indian studies reported diverse communication channels/sources in the different stages of the innovation decision process. Personal cosmopolite channels were reported important (Kapoor, 1966; Sarkar, 1981; Singh, 1989), followed by personal localite channel

(Ernest, 1973; Sawant *et al.*, 1979) at the knowledge stage. Similarly, importance of these channels/sources were recognized at persuasion stage (Reddy and Singh, 1977; Babu and Sinha, 1985) and at decision making stage (Rogers and Pitzer, 1960; Sinha and Prasad, 1966; Singh, 1989) of innovation decision process. It was also highlighted that for technically complex information, cosmopolite channels were important (Sandhu, 1967; Singh and Jha, 1971). Studies in later eighties showed the influence of mass media on Indian farmers as a source of information (Tyagi and Sohal, 1984; Bhagat and Mathur, 1985; Gupta, 1991).

Various research studies conducted in India maintained that agricultural communication, by and large, followed a system approach (Singh, 1988). This consisted of three distinctive subsystems; the research system, the extension system and the client system. This approach, however, was highly top down in nature and suggested that the function of client system were the "adoption of innovation" and "feedback".

Jones (1990) used a system perspective to apply a holistic approach to a relevant defined whole and the activities within it. He emphasized the flexibility of the system as a concept. Similarly, studies by Singh (1970), Havelock (1971), Nagell (1980), Swanson *et al.* (1984), Ambastha (1986), Samantha (1990), Delman (1991) and Rolling and Engel (1992) unequivocally agreed that a system perspective was helpful to a discussion of information transfer in agriculture. In the consonance with above authors, Malik (1993), Bharati (1993), Singh (1994) and Gupta (1998) studied the communication among research, extension and farmers with systems perspective.

2.2 Linkages Between Research, Extension and Farmers

2.2.1 Linkages Between Research and Extension

Most of the early attempts to understand the linkages between research and extension were in the form of communication linkage. This envisaged the information flow pattern between these two in terms of information input, information processing and information output pattern (Lionberger and Chang, 1970; Ambastha and Singh, 1977; Sanoria and Singh, 1978; Reddy, 1984; Verma, 1987; Alzahrwal, 1992; Singh, 1998). At the same time, however, authors like Lionberger and Chang (1970) and Ganorkar and Khonde (1979), Reddy (1984) highlighted the media and methods used by researchers and extensionists to communicate with each other. They reported that personal contact, training, meeting, extension activities, literature and radio were some of those media and methods. Such type of studies lacked linkages perspective and hence the extent of mutual and reciprocal interaction between research and extension could not be explored and worked out. However, to strengthen the process of design and delivery of agricultural technology, the need for strong reciprocal linkages between research and extension was greatly recognized (Jain, 1970; Azad, 1975; Rao and Sohal, 1978). From late eighties, there have been growing awareness at global fora to strengthen the research and extension linkage to a maximum possible extent. Seegers and Kaimowitz (1989), after reviewing the survey of research extension link in 18 countries concluded that although extension feedback to research is more common in higher resources system, extension workers are also the main source of researchable ideas in any country. Based on the study of Souder (1980) on impact of linkage in product development by the research and development projects in USA, Kaimowitz (1990) derived two key lessons applicable for any research managers. These are (a) vitality of strong link with extension workers and farmers for technology development and delivery, and (b) according high priority to solve linkage related problems. Rolling (1989) and Merill Sands *et al.* (1990) also held the similar opinion.

At the national level, it was found that the linkage between CSIR and ICAR at higher level and coordination between district level information delivery system for agriculture, industry, cottage, etc. was quite weak (Gupta, 1991). Singh *et al.* (1991) stated that in the training and visit system, there is well defined linkage structure at all level, *i.e.*, central, state, regional and district. It was, however, a matter of great concern that these linkage points are not functioning as anticipated in the programmes in some states. Rahiman (1991) also revealed the presence of appreciable structural linkage between extension and research system, but the functional linkage was not that encouraging.

Malik (1993) found that the linkage between research and extension personnel of various levels and cadre was inadequate in Haryana owing to number of limiting factors. Similarly, in a comprehensive study of dairy development in Haryana with a systems perspective, Bharati (1993) noted that there was considerable gap between the means of desired interaction and actual interaction between research and extension personnel.

Singh (1994) reported that most of the research and extension personnel (about 63%) expressed rare (15%) to very low (5%) extent of functional linkage between them. His study also revealed that there was relatively better linkage between research and extension in agriculture (45%) than in livestock (20%). On the various parameters of functional linkage, respondents from college of agriculture and SDA felt a good linkage on communication, decision making and training in that order. However, very poor communication linkage between research and extension was noted in College of Veterinary and SDAH. On the other parameters, *viz.*, planning, team work, decision making and supply and services, the functional linkage was poor. In a study to understand the dairy knowledge and information system (DKIS) in the Karnal district of Haryana, Gupta (1998) reported weak to absent linkage between research and extension within NDRI, Karnal and between NDRI and SDAH.

With respect to the effective functional linkage between research and extension, several researchers have argued about the effectiveness of functional linkage *vis-a-vis* their organizational arrangements. Some of them have supported to keep research and extension in the same organization, while others have advocated for their separation. Bourgeosis (1989) have cited two major advantages of merging research and extension in one organization. First, organizational proximity can promote a shared goal and facilitate communication and collaboration between them. Second, merging assume to increase efficiency. Kessaba (1989), Pineiro (1989) and Antholt (1990) also reiterated the similar arguments. Trents (1989), however, on the basis of his experience in Gambia, said that merging two groups in one institution/unit is no guarantee that they will work together. More clearly, merging at the institutional or team level would not eliminate the need to manage links, it can make the need still more acute. In the similar orientation, Eponou (1993) found that integration of research and transfer was rated as fair to good in only one of the five subsystems, where two groups were merged. On the other hand, Kaimowitz (1989) opined that research and technology transfer in one organization create stronger link between them and it should be considered only under certain situation. On the basis of comparative analysis of two institutions in Colombia, he identified following requirements:

- ❑ Research and technology transfer share a common goal and sharply focused area of concern,
- ❑ Human, financial and managerial resources are adequate to support linkage activities, and
- ❑ Managers have a strong commitment to make the two group work together without politicizing them.

Kaimowitz (1989) further argued that where these conditions do not hold, it is probably better to keep the two activities separate.

2.2.2 Linkages Between Extension and Farmers

This part of review deals with the earlier works done with respect to reciprocal interaction between extension and farmers.

The work by Aukhouri (1973), Ambastha and Singh (1976) and Sridhar and Reddy (1977) revealed that extension personnel used office calls, farm and home visits, group meeting, training and demonstrations as the main channels/media of interaction with farmers. A similar kind of findings was also reported by Reddy (1984), Babu and Sinha (1985), Verma, 1987) and Rao (1992). Similarly, Malik (1993) and Singh (1998) studied the interaction and communication dynamics between extension and farmers, and reported a number of media utilized by them. Various studies only indicated the media/sources commonly utilized by farmers in order to seek information related to improved farming. The most important sources and methods used were extension workers, progressive farmers and friends (Ganorkar and Bhugal, 1978; Ganorkar and Khonde, 1979); opinion leaders, veterinary officers (Sawant *et al.*, 1979; Reddy, 1984) and group discussion, cosmopolite and mass media sources, and farmer to-farmer-approach (Singh, 1989; Hall, 1992; Sihag and Grover, 1992; Malik, 1993).

These literature indicate that extension-farmer continuum was studied in isolation and researchers seldom attempted to study interaction between these two actors of development with linkage perspective. In yet another improved version of investigation, Gupta (1998) studied linkage of farmers and extension personnel from the view points of joint diagnosis of problem to design and implementation of solution together. The author, however, could not find any encouraging strength of linkage between two. Extension and farmers interaction was observed to be top down in nature and almost nil participation of clients was noted in development activities. The similar findings were also reported by Sharma and Rao (1998).

2.2.3 Linkages Between Research and Farmers

In order to develop synchronous perception about meaningful research, the scientists and the farmers have to develop a systematic interaction on a symbiotic basis. Such a system may ensure more meaningful research which would have direct bearing on farmers economy (Pant, 1994). In the similar vein, Rolling (1989) advocated the direct links of research with farmers in order to ensure that research focuses on priority needs and problems. He further argued that this reversal is essential if applied research is to produce the technology required to fuel agricultural development. In further elaboration, Biggs (1989) identified four distinct modes of farmers participation in research and these were contract participation, consultative participation, collaborative participation and collegiate participation. Gadewar and Ingle (1993) also supported these modes of farmers participation. Merill Sands *et al.* (1990), however, were doubtful that involving farmers in research is not easy. The farmers' participation was more difficult than initially expected. Involving farmers actively in priority setting, planning and review of research were challenging to institutionalize. Yet, Rolling (1989) and Ashby (1990) positively argued that the full benefits of participation are gained only when farmers have active role in setting the course of action. Investigating the absence of linkage between research and farmers, Pickering (1985) found that it was the richer farmers who benefitted from improved recommended technologies, since recommendations based on optimal returns to particular enterprise were often too difficult or risky for a small farmer to apply. Moreover, lack of such feedback also made the research system unresponsive to resource poor farmers' problems.

Regarding the empirical evidence from Indian context, sporadic reports are available. The works of Ambastha and Singh (1976), Sridhar and Reddy (1977) and Ganorkar and Khonde (1979) revealed that farm and home visits, demonstration, group meeting and training were some of the methods adopted by the researchers to communicate with farmers. Similarly, Chennegowda (1983), Verma (1987), Alzahrwal (1992) and Bharati (1993) found that field days, office calls, extension meetings and training were the important modes of interaction of researchers with farmers.

An indepth probe was done by Gupta (1998) to understand the farmers' participation in design, implementation and evaluation of dairy production researches. The author, however, came out with disappointing scenario and despite the existence of mechanism like FSR/E, farmers participation in the research was found negligible.

The literature reviewed in above sub heads clearly indicated that research-extension-farmers linkages were studied in terms of several mutually exclusive continuum. Haverkort and Rolling (1984), however, stressed that agricultural development requires a "mix of conditions". Haverkort *et al.*

(1988) further advocated for participatory technology development in which farmers, researchers and extension workers cooperate in the effort to improve technology. They also argued that the target population should be involved in policy formulation and decision making right from beginning. Samanta (1991) suggested that farmers group and field workshop could be few useful way to help elicit farmers' ideas to improve communication and foster local initiative.

Discussing about Korean perspective, Cho (1996) highlighted the systematic participation of extension specialists in research through institutionalized meetings and others alike forum. Similarly, Balakrishna (1997) strongly advocated for the mutual and reciprocal interaction among researchers, extension personnel and farmers. In the present investigation, efforts have been directed to assess the strength of functional linkage between and among these three actors of development on various identified parameters of linkage.

2.3 Variables/Parameters for Measuring Linkages Between Research, Extension and Farmers

As Prasad and Reddy (1991) has rightly pointed out the linkage being intangible in nature, measuring the same is often difficult. During late eighties and nineties, however, some of development administrators and research managers have attempted to identify the parameters/indicators/variables on which linkage strength could be assessed. Some of the well recognized typology of linkage have been deliberated in the theoretical orientation of this book. In the present subhead, hence, parameters used by earlier researchers have been reviewed.

The works done during seventies to mid eighties, researchers like Ambastha and Singh (1977), Reddy (1984) and Verma (1987) and others studied linkage using single parameter, *i.e.*, communication, and this was operationalised as the media/channel/sources used by researchers, extension personnel and farmers to exchange, disseminate and receive the information. Even in early nineties, Malik (1993) and Bharati (1993) studied

Even in early nineties, Malik (1993) and Bharati (1993) studied the linkage on only parameter of communication, yet in a bit improved version. The global scenario, however, was different. In order to strengthen the linkage between research and technology transfer, ISNAR conducted studies in nine developing nations. Kaimowitz (1987), during these studies, identified common areas of functions between research and technology transfer in terms of communication, training and services, input feedback, and testing and adaptive researches.

Eponou (1993) said that the parameters like planning and review, collaborative activity, resource exchange, knowledge dissemination, feedback and coordination should be involved in assessing the linkage between research and extension. Similarly, Singh (1994) measured the extent of linkage between scientific and extension personnel on the indicators like planning, decision making, implementation, evaluation, communication, team work, supply and services, and training in his study conducted in U.P. Nawab *et al.* (1995) in their study on linkage among research, extension and farmers in Pakistan used the indicators like communication, participation in research and extension activities and feedback as the important parameters to link them with each other. Gupta (1998) studied linkage among research, extension and farmers on the parameters like joint identification, planning and designing, and implementation of research and extension activities.

The indicators utilized for present investigation have been identified after thorough review of earlier works and in consultation with the experts. These indicators are aptly elaborated in the subsequent chapter.

3 Factors Affecting the Strength of Research-Extension-Farmers Linkages

There is no single recipe for effective links between agricultural research institutions and technology users. Each institution operates within a specific content over which managers often have little control. However, to make good decisions about links, they need to diagnose the factors present in particular setting. The key contextual factors which a manager should consider fall into four main categories policy factors, resource factors, technical factors and organizational factors (Merrill Sands and Kaimowitz, 1990).

In the present subhead, related literatures have been synthesized under the factor head, *viz.*, organizational, psychological and external factors which are affecting the strength of linkage or could be hypothesized to affect the same.

3.1 Organizational Factors

Axinn and Thorat (1972) found that the extent of linkage was related to the extent of knowledge gap between situations/agencies/systems. Cernea (1981) stressed for more decentralized extension system and argued that it will foster efficient linkage of extension with farmers and researchers. Whereas, Hyami and Rattan (1983) identified resource endowment and policy support, both from organization and government, as the factors affecting the linkage between research and extension.

With particular reference to India, Axinn (1991) said that both the size factor and extent of decentralization are relevant for the nature of linkage required. However, Goals of the organization is yet another determinant of successful linkage. The organizational goals are the objectives that the organization, as a whole, is trying to achieve and the orientation towards goals decides the success or failure of the organization (Khandwalla, 1977). Huli (1989) found that profit orientation of the milk plant had negative influence on the performance of plant. Similarly, Sharma (1994) reported considerable agreement in the ranking of goals by the managers of high and low performing dairy development projects.

Sharma and Motilal (1971) reported that open organizational climate was conducive to the attainment of higher level of achievements and satisfaction of the staff. Prakasam *et. al.* (1979) found that organizational climate had a significant bearing upon the performance of bank employees. According to Sinha (1980), authoritarian and bureaucratic climate were inversely related while rest of the climatic factors were positively related with efficiency rating. Huli (1989) reported that the organizational climate as perceived by the top executive was noticed to be bureaucratic while the functional executives largely perceived affiliation and dependency climate.

Bhattacharya and Talukdar (1996) found "dependency" as the most dominant motivational climate in the gram sevak training centres of the North Eastern region of India. Contradicting to this, however, Singh (1997) reported "extension" as the dominant organizational climate component in the non governmental organizations functioning in Bihar for tribal development. With linkage perspective, Blok and Seegers (1988) found that in many developing countries, the agricultural knowledge information system is administratively divided into research and extension directorate, department or division. This administrative set up might or might not create and adequate structural arrangement. They, however, reiterated that what is important to distinguish the administrative set up from the knowledge and information system.

Though the variables like goals of the organization and organizational climates have not been empirically reported to affect the linkages between research and extension, albeit, linkage being the inherent phenomena of research and extension organization, these variables, hence, could be hypothesized to affect the strength of linkage.

3.2 Personal and Psychological Factors

Ambastha (1980) found that research extension personnel contact span had positive and significant correlation with cadre, education and service experience. He also found that the same variables significantly co-varied with researcher farmers contact span. Ambastha and Singh (1979) reported that extension personnel farmers contact span had significant and negative association with cadre and job commitment and positive correlation with dedication. With respect to development of the organization, Stephenson (1963) and Nakkiran (1968) reported that the attitude of the members, managing committee and staff of the cooperative societies was the important determinant of success of organization. Krishnaraj (1981) also found that the attitude of managerial system had positive and significant relation with efficiency of organization. In another study, Sharma (1994) reported the presence of more number of employees with favourable attitude in good performing organization than the poor one. Singh (1994) found that the variables like age, education and training received were positively and significantly associated with the extent of linkage between research and extension. However, professional experience, achievement motivation and value orientation were non significantly correlated with the same. Based on the experience, author suggested to include the variables like job satisfaction, organizational climate, employee's morale, etc. in future course of such investigations. Bharati (1993) found that cadre was the important variable influencing the interaction of scientists with extension functionaries. He also reported that caste, herd size, extension contact and economic motivation of the farmers were some of the variables affecting their interaction with extension functionaries. No literature could be traced which would indicate the variables affecting the mutual interaction/linkages between extension and farmers and researchers and farmers objectively.

3.3 External Factors

Most discussions of the institutions concerned with agricultural technology focus on their internal dynamics, with little reference to the broader social relation in which they are immersed. This gives the impression that both technology development and technology transfer are autonomous processes that obey their own internal laws of development, independent of their social context (Burmeister, 1985).

Lawrence and Lorsch (1969) propounded the contingency theory in business and stressed that the external environment of an organization is an important factor determining its internal dynamics. Based on this hypothesis, Hayani and Rattan (1971) argued that the pattern of technological change in agriculture is influenced by national resource endowments through the interaction among farmers, researchers and administrators in the political arena. During the late seventies, it was organized that how and when different interest groups start to organize themselves to influence research and extension priorities (Guttman, 1978; deJanvry and LeVeen, 1983; Pineiro and Trigo, 1983; Huffman and McNulty, 1985).

Following the work of Heaver (1981) on the interaction between bureaucratic politics and external incentives, Sims and Leonard (1989) developed a comprehensive theory for the specific case of agricultural technology in developing countries. They coined the term "default incentives" to describe the behaviour of researchers and extension workers in the absence of pressure from outside their institutions. In order to validate the above hypothesis, Kaimowitz (1989) carried out the case studies in seven developing countries. He conceptualized four groups that constituted the principal source of external pressure for technology development and delivery. These were : National policy makers (government policy), foreign agencies, farmers and private sector. In the similar line, Kaimowitz *et. al.* (1990) stipulated that positive external pressure was necessary for any agricultural technology system to be responsive to farmers' needs. The present investigation also included the above four elements as the sources of external pressure for research and extension personnel operating under the selected organizational systems and affecting their linkages.

4 Identified Constraints in Maintaining Linkages among Research, Extension and Farmers

A number of reasons have been given to explain the gap between research and extension. For example, separate institutional housing decreases the opportunities to work together. In a study from developing countries, extension managers ranked the lack of links with research as seventh in order of importance out of nine problems they faced (Sigman and Swanson, 1984; Balaguru and Rajagopalan, 1986). Administrative procedure that creates separate work plans, as well as the failure to budget enough money to conduct joint field work and training are another part of the problem (Coulter, 1983). Similarly, Bennett (1988) found that administrative distinction between research and extension leaves an information gap which can not be abridged easily by linkage mechanism.

Attitudinal problems, arising from the socio-economic gap that sometimes exists between the two sets of workers may also contribute to research extension gap. Extension workers are often less well paid, less educated, and work at less prestigious jobs than researchers and feel they are under valued, while researchers might not see the need to work with extension staff and thus, remain to themselves (Coulter, 1983). Extension workers feel that researchers are not part of the "real world", staying only in their laboratories and ignoring the applied aspect of research (Compton, 1984).

The problem of establishing effective research extension linkages appears frequently in the literature. Cummings (1981) noted that R&E have traditionally been separate organizations with no functional linkages. Fernandez (1981) reported on a 1975-76 survey of eight extension services and six research institutes in Central and South America that revealed a wide gap between the two entities. Mosher (1978) called the relationship between R&E organizations a "quarrel", while Maalouf (1983) identified the lack of official linkages as the most regrettable situation, "noting that with a few exceptions" not a single model of this relationship has been found effective. Snyder (1988) also summed up that the poor inter-organizational relation between extension agency and research organization almost guarantee that research result will not reach farmers.

Too often analysis of this linkage has revealed the weakness of the research and development system in agriculture research system. The lack of understanding of each other's role, nature of taste, its constraints has caused a widening empathy and communication gap. Sivaraman (1978) in a paper on the agriculture research system in India noted that a class and caste tendency permeates much of the research and extension system and that field workers and those employed in applied research have lower status, pay and motivation.

Administrative or organizational integration of research and extension under one authority function has found to cause professional conflict, the only feasible exercise is coordinating these two units through an overall council approach. A better insight of the major issues must be interrelated, namely basic with applied or adaptive research and the role of scientists with extension personnel is the research design and development process.

As it has been very rightly stated by Bunting (1983) for development, the most significant, but usually the least effective part of the knowledge system is the extension or advisory sector, through which the objectives, potentials and difficulties of producers can be known, and by which both the products of new research, and accumulated experiences of the past research and practice are conveyed to them. This part of the system is too often conceived as one way stream, through which the "technology" developed by research workers seen as the lead agents, is transferred by extension workers to the expectant producers. This flow often fails, sometimes because the technology is inappropriate, often a considerable time lag exists between the availability of new research findings and their application by the farming community, extension system can work effectively with farmers only if it can offer relevant innovative information. Lack of effective linkage affects the quality of research which is not sufficiently oriented towards the need, problem, resource constraints of the farmers. The important reverse flow, which should be the leading stage in the whole system, is often omitted, neglected or even held in contempt.

Jain (1985) made an effort to analyze some of the major constraints on the effective establishment of research-extension linkages in India. Some of the crucial issues that deserve attention and debate are :

- The lack of an effective role of the State Department of Agriculture (SDA) in identifying the specific production problems of a zone.
- The ineffective utilization of the opportunities for getting feedback through programmes, which are implemented by scientists.
- The gap between the technology and that being accepted by farmers, which has to be bridged by developing stronger links between scientists and official of the SDA in large scale testing and demonstration of the generated technology.
- The lack of formal links between research and extension activities, which need to be supplemented by frequent joint visits and informal discussions on farmers' fields.

He further stated that despite all efforts, however, research and extension linkages are still quite weak in India for three main reasons:

- Scientists at the Agricultural Universities and official of the SDA are reluctant to accept changes in concepts and procedures.
- Although the programmes of the SDA and the Universities bring officials and scientists together, they

often have no real involvement with or appreciation for each other.

- Although the physical infrastructure may be built up at the regional level, decentralization of the administration and management of research and extension is a slow process because of insufficient institutional freedom, competency, and will.

Seegars (1990) also analyzed the cause of poor links between research and extension and he observed that there were differences in the background of research and extension personnel, lack of fit between the extension service and the regional research centre and the expiration of foreign funding.

Singh *et al.* (1991) studied constraints operating in forward and backward linkages of monthly workshop of T&V extension system. They classified and studied into various categories, *i.e.*, training constraints, supply and infrastructural constraints, organizational incentives, coordination, visits, administrative, and budget and financial constraints. Singh (1994) reported constraints in terms of lack of communication activities, lack of adequate funds, lack of incentives and motivation and lack of new technology which affected research and extension interaction.

Similarly, Malik (1993) identified following first five constraints as the major limiting factors in linking extension workers with the farmers:

(*i*) Shortage of funds for contingency and other allowances.

(*ii*) Lack of audio-visual aids.

(*iii*) Difficulty in understanding required touring for want of vehicle.

(*iv*) Problem of coordinating veterinary and animal science activities at KGKs and University level.

(*v*) Lack of interest of work.

Nevertheless, organizational factors, extension techniques employed, infrastructural problems, inputs inadequacy, and social barriers were some of the constraints impeding the effective interaction of extension personnel with the farmers (Bharati, 1993).

Eponou (1996) identified following constraints affecting research-farmers linkage :

(*i*) Lack of explicit linkage policies because of their research strategies.

(*ii*) Lack of awareness.

(*iii*) Perceived transaction of cost and time.

In the similar line, Wuyts (1996) suggested that the research and farmers' organization have the potential to make relevant technologies, but the potential can not be realized unless the associated problems are adequately addressed.

While analysing the planning of research with farmers at five locations in South Africa with a view to develop appropriate technology, Fischer *et al.* (1996) observed the common difficulties encountered included farmers' expectation of free inputs, adherence of farmers and facilitators with TOT mentality and loss of interest by the farmers after planning exercise. Working on similar lines, Farrington (1997) made review of a decade of work on farmers' participation in research and extension. He found that lack of clarity in the objectives of kinds of participation, their mode of client orientation and the various roles of different organizations in promoting participation were the major problems affecting farmers' participation in research and extension.

The perusal of above reviewed literature done in this and previous chapters indicated that most of the works done in past included mainly communication as the main linkage parameter. This highlighted the media and sources used by research and extension personnel to contact with each other and with farmers. Very few workers (Singh, 1994; Gupta, 1998) have included entire dimensions of linkages between research, extension and farmers with systems perspective. In the present investigation, hence, attempt was made to measure the strength of functional linkage between research, extension and farmers in dairying under various organizational systems. The study also attempted to bridge the lacunae by including various other parameters of functional linkage like collaborative professional activities, planning and decision making, implementation and evaluation, training, and supply and services. Further, the study also contemplated to measure the linkage quantitatively and identify the factors affecting the strength of linkage objectively, which were absent in past researches. Hence, the present investigation was, perhaps, a maiden attempt to understand and analyze the research, extension and farmers linkage in a comprehensive manner.

5 Justification for Empirical Probe on Research-Extension-Farmer Linkage in Dairying

In Indian scenario, where agriculture sector sustains the livelihood of majority of the population, farming has the direct bearing on the living condition of rural people. Indian farming is an economic symbiosis between crop and animal husbandry. Dairying, a big constituent of animal husbandry, plays pivotal role in mixed farming of rural areas and provides additional income and productive employment to farm families. India is bestowed with the largest stock of milch animals (about 100 million bovines), which also include the finest breed of buffalo. Since the inception of planned development of agriculture and allied sectors, dairying has received a considerable investment and the network of infrastructure has been created in terms of research institutions and development departments for the ultimate success of dairying. These efforts have yielded the dividends. The milk production has increased about 3.5 folds from the base year of 1951. India is projected to be largest milk producer country in a year to come (Dairy India, 1997). Contrary to this, the per unit production (average 1.5 litres/animal/day) is far below when the global figure is taken into consideration. Further, the acceptance of the recommended dairy production technologies is low to medium level as reported by a number of authors (Gupta, 1974; Mahipal and Kherde, 1989; Sah, 1996). The main reasons presumed could be the non availability of more appropriate technology (Das, 1996; Hansara, 1996) *vis-a-vis* ineffective delivery of the existing technologies (Baweja, 1974) available with the developmental departments. Disseminating the dairy production technology is essentially the responsibility of the field extension agencies. The farm technology has to be transferred from the source of its origin to the ultimate users in such a way that the receiver could know, comprehend, accept and act upon it efficiently. Guba, way back in 1968 had pointed out a tremendous gap between knowledge production and knowledge utilization. After a year, in 1969, All India Seminar on Research in Extension concluded that "a lot of information is generated in agricultural universities, colleges and research institutions. The quickness with which the information is transmitted to the change agent and other prospective clients will determine the success of innovations and programmes".

Since the introduction of National Extension Service in 1953 and subsequent establishment of agricultural universities in almost all of the States, efforts are being directed to channelise the scientific information to the farmers which depends on closed and reciprocal interaction among the three distinct systems, *i.e.*, research, extension and clients. Any detailed consideration of the dissemination and utilization of knowledge must apparently focus on the question of linking roles of extension agency with the technology users and technology producers (by offering research feedback to amend the technology for making them more appropriate). Lionberger and Chang (1970) had rightly specified that the more is the meaningful interaction within and between these three systems, the faster will be the process of modernization of agriculture. It is in this context, hence, the interdependence and linkage among research, extension and farmers becomes more relevant and a subject of investigation.

In the further evolution to strengthen the research and extension infrastructure and capacity, Indian Council of Agricultural Research (ICAR) introduced National Agricultural Research Project (NARP) in 1979. The main aim of NARP was to upgrade and strengthen the regional research capacity of the SAUs. The project has in built mechanism for speedy transfer of technology. As a third effort during 1983, based on the model of NARP and assisted by World Bank, National Agricultural Extension Programme (NAEP) was launched to strengthen the T&V System and, ultimately, for establishing better linkages between researchers and extension workers. Both NARP and NAEP intended to do their jobs efficiently. However, several mid term and terminal evaluation of these projects have revealed that though various fora existed, their impact was not operationally seen at various level (Kaurani, 1995). As a sequel to this, a two days National Seminar was held in 1993 to formulate National Agricultural Technology Project (NATP). The major outcome of this Seminar recommended that although research and extension systems are two different systems, they should be considered as a continuum representing a symbiotic relationship. The project also emphasized greater participation of farmers and involvement of private sector in research and extension services, and could be brought under a Joint Research Extension Management Board with representation from research, extension, NGOs, farmers and private sector organizations. Under the umbrella of NATP, World Bank funded programme, named Technology Assessment and Refinement through Institute Village Linkage Programme (IVLP) was launched in the year 1995 which has basic emphasis on research and farmers' linkage. The above discussion revealed the importance and the mechanism that how the research, extension and farmers' linkage evolved over the year.

Despite the universal consensus regarding the importance and need of linkage within or between the units, only few empirical studies are available on this aspect. Most of the studies have emphasized the communication pattern among information generation, information dissemination and information utilization subsystems. Little care has been taken to ascertain the functional interaction of these three entities on several parameters other than communication. About the empirical probe of linkages, Prasad (1988) has aptly remarked that linkage is a complex phenomena and has complex relationship, basically subjective in nature and at several instance, it takes the form of multi-dimensional subject and multi-institutional approach. Only a handful literature is available in this field which are mostly conceptual and theoretical, and mainly in the terms of suggestions and recommendations for linkages between these three entities. The present study, probably, was a maiden effort to study the linkage in such a comprehensive manner in the field of dairying, research and extension.

5.1 Statement of the Problem

The urgent need for food production for the ever increasing population forced the country to focus attention on agricultural education, research and extension of applicable research findings to the concerned clientele.

Interestingly, the University Education Commission (1949), headed by Dr. S. Radhakrishnan had already emphasized the need for establishing Rural Universities. Further, the 'First Joint Indo American Team on Agricultural Research and Education' (1955) endorsed the recommendations of the Radhakrishnan Commission and recommended for setting up of agricultural universities on the pattern of 'Land Grant Colleges' of the USA. The actual role performance of agricultural universities in agricultural and dairy development is concerned, different agricultural universities have come out with varying outcome of their roles and performance depending upon the State's interest, financial and other limitations. The actual role played by some of the leading agricultural universities in the country, including Chaudhary Charan Singh Haryana Agricultural University, Hisar (henceforth to be referred as HAU) in rural development are as follows :

- Imparting education in the various subjects of agriculture, veterinary science, animal science, home science, etc. so as to generate trained human resource for the scientific development of agriculture, livestock and allied rural sector.
- Generating knowledge and development of appropriate technology through research for providing solutions to the problems confronted in optimising agricultural and livestock production.
- Undertaking location specific agricultural research with the establishment of strong multi-disciplinary regional research stations, so as to develop need based and location-specific technology for the farmers.
- Transferring the applicable and useful research findings, and improved technologies in agriculture and allied fields to the concerned clientele quickly and efficiently.

Apart from HAU, there is another institution in Haryana, operating under the aegis of Indian Council of Agricultural Research (ICAR), named National Dairy Research Institute (NDRI), Karnal. Right from 1955, NDRI is engaged in human resource development, knowlege and technology generation in dairying and their dissemination to the ultimate users. Both, HAU and NDRI embrace the concept of education, research and extension from the same organization. These two organizations constitute the system where research and extension are placed in the same organization.

Yet another organization which is sincerely engaged in livestock and dairy development of the State is the State Department of Animal Husbandry (SDAH). The department is striving hard in dissemination of the dairy production information and technologies to the cattle owners through the network of infrastructure spread throughout the State. Apart from supply of technical inputs and rendering technical services to cattle owners, department also undertake various educational activities to promote dairy development in the State. The SDAH, thus, represents the organization system which is engaged solely in transfer of technology and operates in separation with the research institutions. The detailed accounts of these three organizations/departments working for the cause of dairy development of Haryana is furnished in Annexures I, II and III.

The efficient technology transfer requires inter departmental, intra departmental interaction both within and outside the university and institution. In other words, there should be functional linkage which ought not only be efficient, but also effective. In a Seminar on "Research and Development Linkages and Feedback in Agricultural Development", Jakhar (1993) urged agricultural scientists, extension workers, officials and non official agencies to strengthen the information coordination work at the grassroot level. Similar opinion was reiterated by Chopra (1992). These three organizations having been in operation for several decades, it was considered quite inevitable to have critical analysis of the linkages among research, extension and dairy farmers operating under different organizational systems.

Probing the interaction among the above entities had been the matter of interest to the communication specialists in past also. They have studied communication pattern of the researchers, extension personnel and the farmers under two research traditions, *viz.*, diffusion of innovation and scientific communication. A number of studies from all over the world (Rogers and Shoemaker, 1971) including India (Singh *et al.*, 1973) have been made in diffusion research mainly dealing with flow of information among the farmers. Similarly, the communication behaviour of extension personnel (Patel and Leagans, 1968; Yadava, 1971; Akhouri, 1973; Shete, 1974) and farmers (Singh, 1965; Murthy, 1969; Ernest, 1973) were studied in past, but in isolation. During seventies, however, a number of authors, *viz.*, Lionberger and Chang (1970), Akhouri (1973), Ambastha (1974), Sanoria (1974) and Sridhar and Reddy (1977) attempted to investigate into the effectiveness of information flow from the origin of the innovation to its final adoption by farmers. The similar style of probing the communication pattern among research, extension and farmers continued in eighties (Reddy, 1984). However, in early nineties, a few authors, *viz.*, Bharati (1993), Malik (1993) and Singh (1994) attempted to study the linkages in relatively modified orientation. Still, these studies suffer a serious lacunae in terms of their non comprehensive nature. They scarcely tried to assess the strength of mutual linkages between research extension, extension farmers and research farmers. Only Singh (1994) attempted to measure the extent of functional linkage between research and extension. Study of linkages among these three entities would remain incomplete, if only parameter 'communication' is taken into consideration. Several other indicators, *viz.*, collaborative professional activities, planning and decision making, implementation and evaluation, etc. could be included to make an indepth probe into the intangible subject like linkages. Apart from these, studying the problems of linkages between research and extension would be a matter of interest when both are operating under various organization systems, *i.e.*, when both research and extension are operating in the same organization and when both are working separately.

It is in this context, following questions arise which demand an empirical investigation:

- What are the structural mechanism existing to promote interaction among research, extension and farmers?
- What is the strength of functional linkage between research and extension under various organizational systems?
- What is the strength of functional linkage of farmers and extension, and research and farmers under different extension and research settings?
- What could be the factors (personal, psychological and organizational) predicting the strength of research-extension, extension-farmers and research-farmers linkages?
- Similarly, what are factors (socio-economic, psychological, communication, etc.) predicting the strength of functional linkage of farmers with extension?
- Lastly, what are the bottlenecks impeding the extent of linkages among research, extension and farmers.

In order to find the precise and empirical answer to these research questions of utmost academic and practical utility, the study entitled **"A Study on Linkages Among Research, Extension and Dairy Farmers in Haryana"** was undertaken. The overall objective was to assess the strength of linkage among the three entities, *viz.*, research, extension and dairy farmers.

5.2 Objectives

The specific objectives formulated for the present investigation were :

- To develop the indices and to measure the extent of structural and functional linkages among research, extension and clients.
- To identify the factors influencing the strength of linkage on research-farmer continuum.

- ❑ To delineate the linkage constraints as perceived by researchers, extension personnel and farmers.

5.3 Utility and Scope of Such Probe

The need and importance of effective linkages between or within systems/organizations/institutions are discussed on almost every point of professional discussion at higher level, but little realistic efforts have been made in this direction. Hence, in the process of technology generation and technology transfer, analyses of linkages constitutes an highly important step.

Understanding the linkage mechanism would unravel the nature, mode and itensity of interaction between and among the three entities. The ground reality of the present status of linkages among research, extension personnel and dairy farmers would be ascertained quantitatively. The study would generate information to understand the complex process of linkages.

Findings of the investigation would enable the researcher to make some worthwhile recommendations which could be utilized by the research managers and development practitioners for further management of linkages among these three actors of development.

Study also identify the most significant variables predicting to strength of linkages. These variables, if suitably manipulated by the leader of the organization/department, would result efficient management of linkage activities.

Study also aims to highlight some of the constraints as experienced by research personnel, extension personnel and dairy farmers. Amelioration of these bottlenecks would be helpful in magnifying the strength of linkages.

Study of this kind, hence, assumes maximum scope and utility for the effective management of research and extension activities.

5.4 Limitations of Such Probe

Though, the present study resumes great academic and practical relevance, it has the following major limitations :

- Study suffers from usual limitation of time, money and other resources as it being the student's project.
- The findings are based on the ability and honesty of the respondent in providing their response. Hence, the prejudices and biasedness in their response is not ruled out.
- The study was conducted in particular condition, system and sample, thus the results of the study findings their applicability in the similar set up.
- Although, study included most of the variables relevant for the study, some of more variables may be still missing.

Nevertheless, it is hoped that this study would provide a better conceptual and empirical background to understand the linkage pattern of research, extension and client. The findings of the study would be immensely useful for the planners, policy makers and development managers who are involved in the plight of dairy farming in the State of Haryana.

6 Methodology Adopted

In this chapter are described various research methods and techniques used for the present investigation. These have been discussed under the following sub heads :

6.1 Locale of the Study

6.1.1 Selection of State

The study was conducted in the State of Haryana. Following reasons supported the selection of the State :

(*i*) The State has witnessed a tremendous success in dairy development. According to Dairy India (1997), following points highlight the dairy development in the State :

(*a*) Haryana belongs to Northern dairy region of the country and this region has highest contribution (44.3%) to the national milk production. In Northern region, Haryana shares 13.28 per cent of the total milk production. On national basis, this share is about 6 per cent.

(*b*) The per capita milk production in Haryana is 592 gms/day which is second highest and next only to Punjab (800 gms/day).

(*c*) The State has registered an appreciable growth of 17 per cent in the total breedable bovine population which are in milk. A negative growth of 5.4 per cent in the breedable dry bovine is the indicator of successful dairy development in the State. Overall,

12 per cent growth rate has been witnessed by this State in the total breedable bovines (year 1982, 87).

(*d*) The State Haryana has a total of 78 dairy plants registered under MMPO as on June 1, 1996. This constitute more than 18 per cent of the national figure.

(*ii*) The State has one of the oldest and best Agricultural University (HAU, Hisar) and ICAR Institute (NDRI, Karnal) engaged in research and extension in animal husbandry and dairying.

(*iii*) Researchers' acquaintance with the area was also one of the reasons to select this State for study.

6.1.1.1 Brief Description of the State

The State Haryana is situated in the Northern region of the country. It is flanked by Punjab, Rajasthan, Delhi and Uttar Pradesh. Haryana has semi arid climate in the South West and Gangetic plain environment in the remaining part of the State. Haryana State extends from 27 E 39' to 30 E 55' N latitude and 74 E 27' to 77 E 36' E longitude. It is one of the smallest State of India with an area of about 42 thousand sq.km. The State mostly covers the Indo Gangetic plains and it is divided into six physiographic units. The climate is generally very hot in summer and remarkably cold in winter. Average annual rainfall ranges from 300 mm to 1200 mm. Net irrigated area of the State is 61 per cent. The State has good share of livestock population.

6.1.2 Selection of District

In order to sample, the SDAH functionaries as well as the dairy farmers, the district Karnal was selected purposely. Following reasons supported the purposive selection of this district :

(*i*) The district Karnal has the common area of operation of HAU, Hisar; NDRI, Karnal and SDAH, Karnal.

(*ii*) The ICDP, Karnal under the SDAH is the oldest and best performing ICDP among all in the Haryana State (Sharma, 1994).

(*iii*) The district has one research farm and the Krishi Gyan Kendra (KGK) under HAU, Hisar involved in research and development in agriculture and livestock. The KGK, Karnal is one of the oldest (since 1976) among KGKs in Haryana.

(*iv*) Moreover, not a single study of this kind and coverage had been contemplated in the past.

6.1.2.1 Brief Description of the District

Karnal district lies on the Western bank of the river Yamuna. The river separates Haryana from Uttar Pradesh (U.P.). Karnal district lies between 29 E 09′50" and 29 E 59′42" North latitude and 76 E 31′51" and 77 E 12′45" East longitude. Its height from sea level is between 235 and 252 metres. It is surrounded by Kurukshetra district on its North and North West, Jind district on its South West and Panipat on its South. The climate ranges from dry and hot summers to cold winters. Average annual rainfall is about 44.7 mm. Net irrigated area form about 93.48 per cent of the net cultivated area.

The district Karnal has 11.25 per cent of the total crossbred cows, about 6 per cent of total indigenous cows and 5.46 per cent of the total buffalo population of the State. The district share considerable percentage (about 6%) of the total milk production of the State (Dairy India, 1997).

6.2 Sampling Plan

The study included three sets of sampled respondents Research personnel, Extension personnel and Dairy farmers.

6.2.1 Selection of Research and Extension Organizations/Departments

Since the study was spread over the State of Haryana,

hence, both the organizations named HAU, Hisar and NDRI, Karnal were taken for the present study. Apart from these two, SDAH, Karnal was also included in the study. Both, HAU, Hisar and NDRI, Karnal offered the system of organization, wherein both research and extension were being undertaken. On the other hand, SDAH offered the system of organization where only extension works were being performed. The idea behind selection of such systems of organization was to assess the strength of functional linkage between research and extension personnel when both are placed in same organization (System 1) and when both operated separately (System 2). The brief account of these three selected organizations/departments has been furnished in Annexures I, II and III.

6.2.2 Selection of Research Personnel

Research personnel were selected from both HAU, Hisar and NDRI, Karnal. All the research personnel engaged in applied research in animal production and veterinary were identified in these two organizations. Depending on the availability of research personnel at the time of investigation, a sample of 17 and 15 such personnel were drawn from NDRI and HAU, respectively. Care was taken that these personnel may range from Assistant Professor/Scientist to Professor/Principal Scientist. For the selection of Assistant Professor/Scientist, a criterion of five years research experience was taken into consideration. Thus, a total of 32 research personnel were selected for the present study.

6.2.3 Selection of Extension Personnel

As the study offered to select extension organizations/ departments as per the criteria mentioned in sub head of 4.2.1, three extension systems were selected accordingly. Thus, from HAU, Directorate of Extension (DOE) and field extension wing (mainly KGKs) and from NDRI, transfer of technology wings (KVK, ORP/FSR, IVLP and Division of Dairy Extension) were taken as extension departments. These are operating in close proximity with researchers. Thus, the requirement of System 1 was met in this way.

Similarly, the SDAH which is operating separately from research organizations and is actively involved in transfer of dairy production technologies was selected to meet the condition of System 2.

Depending on the availability of extension personnel at the time of investigation, a sample of 16, 15 and 16 extension personnel were drawn from NDRI, HAU and SDAH, respectively. While selecting these personnel, care was taken that top level, middle level and field level extension personnel may be given their due representation. Thus, a total of 47 extension personnel formed another set of sample for the study.

Table 6.1 : Distribution of the dairy farmers of the selected villages based on their actual population and sample drawn.

Extension Systems	*Name of the Villages*	*Actual Population*				*Sample Size*			
		Small (Upto 3)	*Medium (4-5)*	*Large (> 5)*	*Total*	*Small (Upto 3)*	*Medium (4-5)*	*Large (> 5)*	*Total*
NDRI	Sanghoa	600	1000	400	2000	6	10	4	20
	Kulwehri	136	176	78	390	7	9	4	20
	Kailash	120	120	60	300	8	8	4	20
CCS HAU	Dinger Majra	140	160	100	400	7	8	5	20
	Mahmadpur	65	105	30	200	7	10	3	20
	Kachhwa	400	350	250	400	8	7	5	20
SDAH	Dabri	75	125	50	250	6	10	4	20
	Badauta	105	105	95	300	7	7	6	20
	Gogdipur	175	200	125	500	7	8	5	20
	Total	**1816**	**2341**	**1183**	**5540**	**63**	**77**	**40**	**180**

6.2.4 Selection of Villages and Dairy Farmers

Villages were chosen from the areas of operation of all three types of selected extension systems operating in the district of Karnal. A total of 9 such villages were selected randomly, 3 from each extension system. From each selected village, 20 dairy animal owners representing small (up to 3), medium (4-5) and large (more than 5) farmers were taken by probability proportionate to their size. The classification of dairy farmers into above categories was done on the basis of adult milch animals possessed by them. The technique of cumulative square root was employed for stratification of dairy farmers. The distribution of actual population and sample of dairy farmers of the selected nine villages has been given in Table 6.1. Thus, 180 dairy farmers constituted last set of sample for meeting the requirements of the objectives. The distribution of different types of respondents is shown in Table 6.2.

Table 6.2 : Distribution of different types of respondents taken for the study.

Sl. No.	*Respondents' Categories*	*Taken From*	*Number*
1.	Research personnel	HAU, Hisar	15
		NDRI, Karnal	17
2.	Extension personnel	HAU, Hisar	15
		NDRI, Karnal	16
		SDAH, Karnal	16
3.	Dairy farmers	Nine adopted villages	180
	Total		**259**

6.3 Variables and their Measurements

This subsection deals with the variables pertinent to the present study, their operationalization in the context of the nature of the investigation and their empirical measurements. The selected variables have been broadly categorized into independent variables, relevant variables and dependent variables.

6.3.1 Independent Variables

6.3.1.1 Variables Pertaining to Research and Extension Personnel

(a) *Age* : Age was considered as an important variable affecting all behavioural acts of an individual including

communication and linkage. Age refers to the chronological age of the respondents in year in whole number. The respondents were categorized into young, middle and old group on the basis of Mean $(\bar{x})$ and Standard Deviation (σ).

(b) ***Education*** **:** Education refers to the number of years spent in formal education and academic credential attained by the respondents which was measured by direct questioning. Following categories of respondents were made according to their educational level:

(*i*) Matriculate (*ii*) Diploma

(*iii*) Graduate (*iv*) Post-graduate

(*v*) Doctorate

(c) ***Cadre*** **:** It indicates the relative position of the research and extension personnel in the hierarchy of the organization/department. It was ascertained by direct questioning and score was assigned as per the designation of different categories of respondents. The personnel were then categorized into junior, middle and senior cadre depending upon their position in the hierarchy of the organization/department.

(d) ***Basic Pay*** **:** It refers to the basic pay of research and extension personnel received at the time of data collection. This was done by direct questioning. The respondents were categorized into low, medium and high pay scale group on the basis of Mean and S.D.

(*e*) ***Professional Experience :*** It is the total number of years completed as research or extension personnel in the organization/department. Based on the Mean and S.D. of job experience, the personnel were categorized in low, medium and high.

(*f*) ***In Service Training :*** It was operationally defined as the extent the personnel have received training for their job enrichment. The response was taken by direct questioning about the duration (in months) of the training received. Based on the frequency, respondents were categorized into four groups, *viz.*, no training, one month, three months, five months, eight months and more than eight months training.

(g) ***Family Type :*** This indicate the nature of family, i.e., nuclear or joint. On the basis of frequency of response, personnel were distributed into above two categories.

(h) ***Family Size :*** This refers to the number of individual dwelling under same roof and sharing kitchen together in a household. The personnel were categorized into small medium and large based on Mean and S.D.

(i) ***Family Background :*** It refers to the various types of families on the basis of location and occupation to which individual belongs. The collected data revealed four types of family background, *viz.*, rural (agricultural), rural (non agricultural), urban (agricultural) and urban (non agricultural). Based on the frequency of response, research and extension personnel were distributed in the above categories.

(j) ***Attitude of Personnel :*** Thurstone (1946) defined attitude as "the degree of positive and negative affects associated with some psychological objects like symbol, phrase, slogan, person, institutions, ideal or ideas towards which people can differ in varying degrees". For the present purpose, attitude was operationalized as Research and Extension personnels' positive or negative affect associated with the organizations/departments. By using the scale as mentioned in Table 6.3, the attitude of personnel was measured towards the aspects like work, working conditions, co-workers, supervisors and management in the organization concerned. A set of 15 statements was given to the individual respondent (Annexure IV). The scoring was done according to a prepared key indicating a positive, negative and neutral response. The algebraic sum of the scores for the items on each attribute, and the same of the scores on different attributes gave an indication of overall attitude. Further, on the basis of Mean and S.D. of these scores, personnel were categorized into favourable (positive), indifferent (neutral) and unfavourable (negative) attitude towards the organizations/departments on the selected dimensions of attitude.

(*k*) ***Achievement Motivation:*** Achievement motivation was conceived as the psycho-personality variable which empels the individual to strive for success for its own sake rather than anticipating the concrete rewards.

The scale employed Singh (1974), to measure the achievement motivation contained six items, out of which two indicate negative and four negative response. The scoring pattern was reversed for the negative statements. Based on the Mean and S.D. of the obtained scores, personnel were categorized into low, medium and high achievement motivation.

(*l*) ***Value Orientation :*** Values are defined as "Modes of organizing conduct meaningful, affectively invested pattern and principles that guides to achieve". In the present study, value orientation of research and extension personnel were studied in respect of two dimensions: Localite Cosmopolite (LoCo) and External Conformity Individualism (EI). The scale developed by Murthy (1969) was used after suitable modification. The scale contained both positive and negative statement. Response from the personnel were ascertained on five point strongly agree-strongly disagree continuum and scores were assigned accordingly. For negative statement, scoring was reversed. The response given individual respondents on each item were summed up and overall Mean and S.D. were calculated. The respondents were categorized into group of low, medium and high value orientation on the basis of Mean and S.D.

(*m*) ***Job Satisfaction :*** This was operationalized as the verbal expression of the research and extension personnels' evaluation of their job. The response was obtained on several parameters of job satisfaction on 5 point continuum satisfied–dissatisfied continuum. The score for negative items was accorded in reverse order than the positive items. Based on the response, the scores were summed up and averaged. Then S.D. was worked out and on the basis of Mean and S.D., personnel were categorized in low, medium and high category.

(*n*) ***Morale*** : For the present study, morale has been operationalized as per Pestonjee (1973), as the general feeling of personnel based upon their faith in the fairness of the employers' policies, adequacy of immediate leadership, a sense of participation in the organization and an overall belief that organization is worth working for. Since, the original scale of this author was intended to measure the morale of industrial workers. Hence, for the present investigation, the scale as modified by Patro (1982) was used to measure the morale of research and extension personnel. Based on the obtained score, respondents were categorized into low, medium and high, after computing the Mean and S.D.

(*o*) ***Perception of Management*** : The management of the selected organizations/departments was studied through the latent variable as the perception of the personnel and farmers towards the same. The idea was that the respondents' perception of management of the organizations/departments would have some bearing on the extent of functional linkage among them.

Based on the scores obtained, respondents were categorized into four categories, *viz.*, low (poor), medium (average) and high (good) on the basis of Mean and S.D.

(*p*) ***Knowledge Gap*** : Knowledge is defined to include those behaviour and test situations which emphasized the remembering either by recognition recall of ideas, material or phenomena. This variable was used to assess the gap in the level of knowledge of Scientific Dairy Farming Practices (SDFPs) of extension personnel. It was operationalized as the gap between recommendation about technology and actually known to the personnel.

Percentage knowledge gap was computed by using following formula :

$$\text{Knowledge Gap}\,(\%) = \frac{\begin{pmatrix}\text{Total scores}\\ \text{allotted}\end{pmatrix} - \begin{pmatrix}\text{Scores}\\ \text{obtained}\end{pmatrix}}{\text{Total scores allotted}} \times 100$$

Based on percentage knowledge gap, extension personnel were categorized into low, medium and high group using Mean and S.D.

(q) ***Goals :*** The goals were operationalized as the objectives that guide the day-to-day as well as long term decisions of the organizations/departments as accorded by the research and extension personnel. Ten statements related to goals of the organization were given to each respondent. Each statement had to be given marks based on its importance so that the total of all statements adds upto 100 (Annexure IV).

(r) ***Organizational Climate :*** All the components of organization–structure, systems, culture, leader behaviour and psychological need of employees interact with one another and create what can be called as organizational climate. Hallriegal and Slocum (1974) defined organizational climate as "a set of attributes which can be perceived about a particular organization and/or its subsystems deal with their members and environment".

In the present study, organizational climate was studied on the dimensions, *viz.,* inter-personal relationship, problem management, conflict management, communication, decision making, trust, risk taking, leadership, etc. Response from research and extension personnel were ascertained on bi-polar statement on ten point continuum against each dimension. The score thus obtained were summed up, and Mean and S.D. were worked out and categorization (low, medium and high) was done accordingly.

(s) ***External Environment :*** This was operationalized as the sources of external pressure (mainly from policy,

farmers' organization, foreign agencies, private organizations, etc.) constraining the research and extension activities in the selected organizations. Each external element was provided with four choices agree, disagree, uncertain and not relevant (Annexure IV).

6.3.1.2 Variables Pertaining to Dairy Farmers

(a) ***Age*** : Refers to the chronological age (in years) of the dairy farmers on the date of investigation. On the basis of Mean and S.D. of the age, dairy farmers were categorized into young, middle and old age group.

(b) ***Education*** : It refers to the number of years spent in formal schooling. Based on these scores, farmers were categorized into following groups :

Sl. No.	*Category*	*Score*
1.	Illiterate	0
2.	Middle	5
3.	Matriculate	10
4.	Intermediate	12
5.	Graduate	15
6.	Post-graduate	17

(c) ***Family Education Status*** : It refers to the academic qualification of the dairy farmer and his family members. It measured through a schedule developed for the same. Family education score has been computed with the help of following formula :

$$FES = \frac{\Sigma N}{N}$$

where,

FES = Family education score,

Σn = Sum total of the educational score of all the educable members (above six years) of the family, and

N = Total number of family above six years.

On the basis of FES, respondents were categorized into low, medium and high group using Mean and S.D.

(d) ***Family Type*** : This indicates the nature of family, *i.e.*, nuclear or joint. On the basis of frequency of response, dairy farmers were distributed into above two categories.

(e) ***Family Size*** : It refers to the number of individuals dwelling under the same roof and sharing kitchen together in a household. It was measured by enumerating one score to each family members. Dairy farmers were categorized into small, medium and high group on the basis of Mean and S.D.

(f) ***Occupation*** : It refers to the main source of livelihood of the dairy farmers. Farmers' response about their primary and secondary occupation was taken and they distributed in the following occupation category based on their frequency of response :

Sl. No.	*Occupation*	*Score*
1.	Agriculture	1
2.	Dairying	2
3.	Business	3
4.	Contract labour	4
5.	Service	5
6.	Others	6

(g) ***Caste*** : It refers to the face or community to which a farmer belongs. The respondents were categorized into three caste groups as upper caste, backward caste and scheduled caste depending upon their affiliation.

(h) ***Land Holding*** : It refers to the area of land being cultivated by the dairy farmers. It was measured by direct questioning and the respondents were categorized on the basis of standard criteria:

Landless	No land
Marginal	Upto 2.5 acres
Small	2.6 to 5.0 acres
Medium	5.1 to 10.0 acres
Large	More than 10.0 acres

(i) *Herd Size :* In present study, it refers to the total number of dairy animals of different age groups possessed by the farmers. It was measured with the help of a schedule and the respondents were categorized into small, medium and large on the basis of Mean and S.D.

(j) *Milk Production, Milk Consumption and Milk Sale :* Milk production was considered as the quantity of milk (in litres) produced in a household, a day prior to the date of enquiry. Milk consumption was the quantity of milk (in litres) consumed per day by the family members. Milk sale refers to the quantity of milk sold (in litres) by the farmers.

By computing Mean and S.D., dairy farmers were categorized in low, medium and high group.

(k) *Social Participation :* It refers to the involvement of an individual farmer in any formal as well as informal social organization/institution as a member. Based on the response, dairy farmers were categorized into no participation and low, medium and high categories on the basis of Mean and S.D.

(l) *Extension Contact :* It refers to the contact made by the dairy farmers with VLDA, stockman, VAS, etc. of the SDAH, cooperative or any other department. A structured schedule was used to measure the frequency of visits by the farmers and personnel to each other. On the basis of obtained score, respondents were classified into low, medium and high using Mean and S.D.

(m) ***Mass Media Score :*** It refers to the degree of utilization of mass media by the dairy farmers. It was measured by a schedule developed by Bhanja (1981). The response was taken on three point continuum of regularly, occasionally and never, and the score of 2, 1, 0 were accorded, respectively. The obtained scores of all the respondents were averaged and S.D. was worked and they were categorized into low, medium and high group.

(n) ***Risk Preferences :*** It is the degree to which farmer is oriented towards risk and uncertainty and has courage to take risk and face the problem of farming.

(o) ***Cosmopolite Localiteness :*** Cosmopoliteness is defined as the tendency of an individual to be in contact with outside his own community, whereas localiteness is the tendency to limit one's contact within the community.

(p) ***Perception of Management :*** It was operationalized as the same done in the subhead of this chapter. Based on the response, farmers were categorized into no perception, low (poor), medium (average) and high (rich) group using Mean and S.D.

6.3.2 Relevant Variables

For the present study, some of variables pertaining to dairy farmers were recognized important but could not be strictly put under either independent or dependent variables. Hence, in the present subheads following such variables have been operationalized.

(a) ***Knowledge Gap :*** Knowledge is defined to include those behaviour and test situations which emphasized the remembering either by recognition or recall of ideas, material or phenomena. For the purpose of present study, knowledge gap was operationalized as the gap between recommendations about scientific dairy farming and actually known to the individual farmer about the same. Percentage knowledge gap was computed by using following formula :

$$\text{Knowledge Gap } (\%) = \frac{\begin{pmatrix}\text{Total scores}\\ \text{allotted}\end{pmatrix} - \begin{pmatrix}\text{Scores}\\ \text{obtained}\end{pmatrix}}{\text{Total scores allotted}} \times 100$$

The farmers respondents were categorized into three groups, *viz.*, low, medium and high using Mean and S.D.

(b) ***Knowledge About the Department :*** It was operationalized as the knowledge of the dairy farmers about the extension departments (NDRI, HAU and SDAH) and their activities. This was measured by using a teacher made type test as developed by Sharma (1994). Based on the response, total knowledge score was worked out and the farmers were categorized into low, medium and high group using Mean and S.D.

(c) ***Gap in the Extent of Adoption :*** Adoption is a decision to make full use of any innovation/practice at the best course of action available. In the present study, gap in the extent of adoption was operationalized as percentage gap between the practices ought to be followed by the farmers and the practices being actually followed by them. For measuring this, an index was developed by following the recommended procedures. The finally selected items (Annexure V) were administered to the dairy farmers and the response was obtained on three points continuum of always, sometimes and never with respective scores of 2, 1 and 0. Based on the response of each item, total adoption score for each farmer was obtained by summing up those scores. Following formula was employed to compute gap in the extent of adoption of SDFPs :

$$\begin{array}{l}\text{Gap in the Extent}\\ \text{of Adoption } (\%)\end{array} = \frac{\begin{pmatrix}\text{Maximum}\\ \text{obtainable}\\ \text{scores}\end{pmatrix} - \begin{pmatrix}\text{Actual}\\ \text{obtained}\\ \text{scores}\end{pmatrix}}{\text{Total scores allotted}} \times 100$$

Based on this value, the dairy farmers were categorized into low, medium and high adoption gap group with the help of Mean and S.D.

Table 6.3 : Variable pertaining to research and extension personnel and their measurement.

Sl. No.	Variables	Measurement
I.	*INDEPENDENT VARIABLES :*	
(A)	**Personal Variables :**	
1.	Age	Direct Questioning
2.	Education	Direct Questioning
3.	Cadre	Direct Questioning
4.	Basic pay	Direct Questioning
5.	Professional experience	Direct Questioning
6.	In service training	Direct Questioning
7.	Family type	Direct Questioning
8.	Family size	Direct Questioning
9.	Family background	Schedule Developed
(B)	**Psychological Variables :**	
10.	Attitude	Hafeez and Subbarya (1974) Scale
11.	Achievement motivation	Singh (1974) Scale
12.	Value orientation	Murthy (1969) Scale
13.	Job satisfaction	Brayfield and Rothe (1951) Scale
14.	Morale	Patro (1982) Scale
15.	Perception of management	Patro (1982) Scale
16.	Knowledge gap	Singh (1994) Index
(C)	**Organizational Variables :**	
17.	Goals	Sharma (1994) Schedule
18.	Organizational climate	Kolbe (1974) Scale
19.	External environment	Schedule Developed
II.	*DEPENDENT VARIABLES :*	
20.	Research - Extension linkage	Index Developed
21.	Extension - Farmers linkage	Index Developed
22.	Research - Farmers linkage	Index Developed

Table 6.4 : Variables Pertaining to the Dairy Farmers and their measurement.

Sl.No.	*Variables*	*Measurement*
I.	**INDEPENDENT VARIABLES :**	
(A)	**Socio Personal Variables :**	
1.	Age	Direct Questioning
2.	Education	Direct Questioning
3.	Family education status	Schedule Developed
4.	Family type	Direct Questioning
5.	Family size	Direct Questioning
6.	Occupation	Schedule Developed
7.	Caste	Direct Questioning
(*B*)	**Socio-Economic Variables :**	
8.	Land holding	Direct Questioning
9.	Herd size	Schedule Developed
10.	Milk production	Schedule Developed
11.	Milk consumption	Schedule Developed
12.	Milk sale	Schedule Developed
(*C*)	**Communication Variables :**	
13.	Social participation	Trivedi (1963) Scale
14.	Extension contact	Singh (1980) Scale
15.	Mass-media exposure	Bhanja (1981) Scale
(*D*)	**Psychological Variables :**	
16.	Risk preference	Supe (1969) Scale
17.	Cosmopolite localite	Singh (1969) Scale
18.	Perception of management	Patro (1982) Scale
II.	**RELEVANT VARIABLES :**	
19.	Knowledge gap	Verma (1993) Index
20.	Gap in extent of adoption	Index Developed
21.	Knowledge about the department	Sharma (1994) Schedule
22.	Source perception	Sarkar (1981) Scale
III.	**DEPENDENT VARIABLES :**	
23.	Farmers - Extension linkage	Index Developed
24.	Farmers - Research linkage	Index Developed

(d) ***Source Perception*** : It refers to the person/place from where the dairy farmers could perceive about the activities and management of the extension organizations/departments. The scoring was done as per Sarkar (1981). Based on the frequency of response, farmers were distributed according to their perception of source.

6.3.3 Constraints in Linkages

The dictionary meaning of 'constraints' is anything that comes in the way of any function, performance and system. In the present investigation, constraints in linkage has been operationalized as the problems which impedes the reciprocal interaction among research personnel, extension personnel and dairy farmers. The constraints were identified by asking open ended questions to the respective respondents (Annexure IX). Based on the response, constraints were explained in terms of frequency and percentage.

6.4 Instruments of Observation

Study had the provision to obtain data on the aspects, *viz.*, socio personal, socio economic, communication, psychological variables of the dairy farmers; personal, psychological and organizational variables for the research and extension personnel. These set of variables were treated as the independent variables and for measuring these variables appropriate schedules were devised and scales were utilized, as mentioned in subhead 6.3.1. Functional linkage of three types, *i.e.*, research-extension, extension-farmers and research-farmers were the dependent variables for the present investigation. These variables were measured by developing the indices as per the procedure mentioned earlier.

6.5 Data Collection

The actual data collection from the intended respondents was done by the investigator himself. Since, data were taken from diverse nature of respondents working at various organizations/department, sufficient time, hence, was devoted in developing rapport with them and in collection of data.

Similarly, while collecting the data from farmers, care was taken to obtain the most valid and authentic information from them. Biasedness in the response was avoided at the level best.

6.6 Statistical Analyses

The collected data were compiled, scored, tabulated and subjected for the following statistical analyses to draw meaningful conclusion:

(*i*) Frequency

(*ii*) Percentage

(*iii*) Average $(\overline{x})$

(*iv*) Standard deviation (σ)

(*v*) Range

(*vi*) Correlational analysis

(*vii*) Regression analysis

(*viii*) Co-efficient of variation (CV)

(*ix*) Least-square analysis of variance (ANOVA).

7 Standardizing the Instrument for Measuring Linkages

Dependent variables for the present study were linkages of different types. Different researchers have defined/used linkages in many ways and types as per the objectives and need of the study. For the purpose of this study, however, linkages was operationalized as the working on functional relationship based on effective communication for a definite purpose. In other words, linkage may be a working or functional relationship for a purposive interaction among the entities, *viz.*, research personnel, extension personnel and dairy farmers. Following two types of linkages were identified and studied.

7.1 Structural Linkage

It refers to the formal interactions which may exist between research and extension organizations/departments and farmers. These interaction include the division of resources, responsibilities for collaboration and different mechanisms that have been created for them to coordinate their activities. This was operationally conceived in the form of structured committee, meetings, discussion forum and alike between the selected organizations/department and farmers.

7.2 Functional Linkage

The functional linkage focus on particular function that need to be performed by the organization/department to link technology generation with technology transfer, technology transfer with technology users and also the technology generation with the users in mutual and reciprocal manner. Based on literature and consultation with judges, different parameters for the functional linkage were identified. These parameters were used to measure the extent of functional linkage among research, extension and farmers. Some of the parameters were common and few were unique for different types of functional linkages, *viz.*, research-extension linkage, extension-farmers linkage and research-farmers linkage. Under the following subheads, the selected parameters of each type of functional linkages have been operationalized and the procedure followed in developing the index of three types have been elaborated.

7.2.1 Research-Extension Linkage

(a) ***Communication :*** Communication is a process by which two or more people exchange ideas, facts, feelings and impressions in a way that each gain a common understanding of the message. It was studied in terms of existing extent of communication between research and extension personnel operating under selected organization systems. This was assessed in terms of the different channels/media used jointly by both the research personnel and extension personnel. The communication included intra as well as inter organizational communication taking place as linkage.

(b) ***Collaborative professional activities :*** This was operationalized as the various activities (field as well as centre activities) which are being performed jointly by the research and extension personnel. It included the activities, *viz.*, joint diagnosis of farmers' problem, identification and evaluation of the solution for such problems, joint formulation and execution any field projects, etc.

(c) ***Planning and decision making :*** Planning refers to the systematic preparation of plans for action either individually or collectively. Similarly, decision making is the process by which the alternatives available to an individual or groups are reduced and finally selected. This was operationalized as the involvement of research and extension personnel in planning and deciding each other's programmes/activities like research projects, adaptive research, trials, survey, camp, campaign, etc.

In addition to these, planning for the resources and planning the monitoring and evaluation of any research/extension activity were also taken as the important items.

(d) ***Implementation and evaluation :*** This was operationalized as the process of arranging the resources and carrying out any research/extension activity jointly. This also included actual monitoring and evaluation of the project/programme and modifying the programme thereupon jointly.

(e) ***Training :*** In this study, the concept of training was used to refers the participation and involvement of research and extension personnel taken from the selected organization system in the process of training programmes for the extension staff and personnel. Relevant items in this respect have been included in the index developed.

(f) ***Supply and services :*** It was used to refer the involvement of research and extension personnel of selected organization systems in assessment of supply of technical inputs, suggestions and offering their services during field supply of those inputs.

7.2.1.1 *Development of Index for Measuring Research-Extension Linkage*

(a) ***Collection of Items :*** All possible items were collected on the selected parameters discussed above. These parameters were selected after consulting the literature and discussion with scientists of extension education of few institutions (NDRI, IARI, IIHR, IVRI, etc.). While selecting these parameters, it was considered that these should reflect/include the whole dimensions of existing linkage between research and extension personnel operating under various organization systems.

Initially, a total of 65 items were framed related to communication, collaborative professional works, planning and decision making, implementation and evaluation, supply and services and training (15, 14, 10, 11, 8 and 7, respectively). The direct and positive type of items were undertaken to measure the linkage.

(b) ***Analysis of items and selection of items :*** A set of items on these selected parameters was sent to judges (40) drawn from various institutions (GBPUA&T, Pantnagar; IVRI, Izatnagar; IARI, New Delhi; UAS, Bangalore; RAJCOVAS, Pondicherry; BAU, Kanke; IIHR, Bangalore; IIM, Ahmedabad) for their response on five point strongly agree-strongly disagree continuum. The scoring was done 5 to 1 for the response continuum for item analysis. These scores were summed up individually for each item on the basis of scores given by individual judge. The mean score was worked out on the basis of acquired sum of scores on each item. The items which were having their mean score above the overall mean score were selected to include in the final index. Thus, in all, 27 items were selected to include in the final index (Annexure VI).

Further, while administering the index on the respondents (Research and Extension personnel), each selected item was weighed on four point continuum always, sometimes, rare and never. This continuum was accorded 3-0 scores, respectively. For calculating the extent of functional linkage, the response for the each parameter was summed up. The overall score (obtained) was thus computed. This score was then divided by the maximum obtainable score to arrive at the percentage extent of linkage score of each respondent. Based on these scores, respondents were categorized into following groups :

Sl. No.	*Linkage Status*	*Score*
1.	No linkage	0
2.	Weak linkage	< 4.42
3.	Moderate linkage	4.42–22.12
4.	Strong linkage	> 22.12

7.2.2 Extension-Farmers Linkage

(a) ***Communication*** : The communication aspect of extension-farmers linkage as perceived by the extension personnel was operationalized in terms of information dissemination behaviour of extension personnel through various media, activities and forum. Whereas for farmers, communication means the information seeking behaviour as well as offering feedback to the department or personnel.

(b) ***Planning and decision making*** : This was operationalized as the working of extension personnel and farmers together to plan and decide the extension programmes/ activities. It included their joint involvement in problem identification to planning of the monitoring and evaluation of any extension activity.

(c) ***Implementation and evaluation*** : This was operationalized as the actual involvement of both the extension personnel and farmers in several activities during the course of actual implementation and evaluation of any extension programme. It included the joint arrangement of different kind of resources, mobilising in the people, monitoring the programme, evaluation of the programme, etc.

(d) ***Supply and services*** : Functional linkage of extension personnel and farmers in supply and services and operationalized as extent to which the technical inputs and services were offered by the personnel and the same were utilized by the farmers.

(e) ***Training*** : Training has been identified as one of the most important intangible investment for the development of any sector in general and dairying in particular. For the present study, linkage in training was operationalized as the extent to which the extension personnel were imparting training to the farmers and the extent to which the farmers were involved in it.

7.2.2.1 *Development of Index for Measuring Extension-Farmers Linkage*

(a) ***Selection of items :*** Firstly, the parameters which could adequately measure the linkage between extension and farmers were identified by consulting the literature and discussing with the scientists of extension education of some institutions (NDRI, IARI, IIHR, IVRI, etc.). These parameters have been properly operationalized in the above sub heads. Further, all possible items were collected on the selected parameters. Initially, a total of 68 items were framed related to communication, planning and decision making, implementation and evaluation, supply and services and training (30, 10, 12, 8 and 8, respectively). The direct and positive type of items were taken to measure linkage.

(b) ***Analysis of items on the selected parameters :*** The set of items on the selected parameters was sent to the judges (40) drawn from various institutions (names of these institutions are mentioned in earlier subhead) for their response on five point strongly agree-strongly disagree continuum. The scoring was done 5 to 1 for the response continuum for item analysis. These scores were summed individually for each item on the basis of scores given by individual judge. The mean score was worked out on the basis of acquired sum of scores on each item. The items which were having their mean score above the overall mean score were selected for the inclusion in the final index. Thus, in all 36 items were selected for final index (Annexure VII).

Further, while administering the index on the respondents (Extension personnel and dairy farmers), each selected item was weighted on four point continuum of always, sometimes, rare and never. This continuum was accorded 30 scores in that order. For calculating the extent of functional linkage, the response for each parameter was summed up. The overall score (obtained) was thus computed. This score was then divided by the maximum obtainable score to arrive at the percentage extent of linkage score of each respondent. The percentage extent of linkage score was computed separately for extension personnel and dairy farmers. Based on these scores, extension personnel and dairy farmers were categorized into following status of functional linkage :

Linkage Status	*Extension Personnel*	*Dairy Farmers*
No linkage	0	0
Weak linkage	< 28.75	< 11.89
Moderate linkage	28.75 – 60.45	11.89 – 31.11
Strong linkage	> 60.45	> 31.11

7.2.3 Research-Farmers Linkage

For understanding the extent of linkage of research personnel and farmers, only one parameter, *i.e.*, communication was recognized. This was operationalized as the degree to which research personnel were offering information related to improved dairy farming through various media, activities, channels, etc. For dairy farmers, it means to degree to which they were seeking information from research personnel and offering feedback to them.

7.2.3.1 Development of an Index for Measuring Research-Farmers Linkage

The procedure as followed in developing the first two types of index was adopted in this case also. Collection of items, analysis of items and selection of items were done as per the similar procedure mentioned earlier. The finally selected items were included for the study (Annexure VIII). The index was administered to the intended respondents and the response was taken on four point continuum of always, sometimes, rare and never and the scores of 3, 2, 1 and 0 were accorded, respectively. Based on the response of each item, overall score was computed. The obtained score was then divided by the maximum obtainable score to arrive at the percentage extent of linkage score. Based on these score, respondents were grouped into following categories :

Sl. No.	*Linkage Status*	*Score*
1.	No linkage	0
2.	Weak linkage	< 8.02
3.	Moderate linkage	8.02–38.16
4.	Strong linkage	> 38.16

8 Empirical Findings on Research-Extension-Farmer Linkage

In this chapter, the results of the study and discussion thereupon are reported under the following heads:

8.1 Structural Linkage Mechanism Between Research, Extension and Dairy Farmers of the Selected Systems

Under this head, some of the structural arrangements as designed by the selected organizations were identified. Their functions, frequency as well 7as status of activities were ascertained from official records and by discussing with the concerned personnel. In the following subheads, findings have been presented and discussed.

8.1.1 Structural Linkage Mechanism Between Research and Extension as Developed by NDRI

The findings contained in Table 8.1 reveal that NDRI has developed a number of structural linkage mechanisms (SLM) for the interaction between research personnel (RP) and extension personnel (EP). Those mechanisms as well as their purpose are contained in the same table. However, it was found, except FSRP and IVLP, frequency of activities of remaining forum ranged from once to twice a year. Further, it was also found that the status of activities of these mechanisms with the personnel from NDRI was regular, but the same with SDAH personnel was nil to very little.

It means that the research personnel from NDRI lacked the platform, where they could interact with the personnel from SDAH. Moreover, very few SLM between the RP and EP from NDRI will have affect on the strength of functional linkages between them.

Table 8.1 : Structural linkage mechanism between research and extension as developed by NDRI

Sl. No.	Linkage Mechanisms	Purpose	Frequency of Activities	Status of Activities with	
				NDRI Personnel	SDAH Personnel
1.	Research Advisory Committee (RAC)	(*a*) Suggests the thrust areas for research programme (*b*) Reviews the research achievements/progress	Twice-Thrice a year	Regular	Very little
2.	Staff Research Council	(*a*) Reviews the new, ongoing and completed research project (*b*) Promotion and planning of multi-disciplinary and multi-locational projects	Twice a year	Regular	Nil
3.	Extension Council	(*a*) Designs policy regarding extension/TOT works of the Institute (*b*) Reviews the extension/TOT activities	Twice a year	Regular	Little
4.	Dairy Husbandry Officers' Workshop (DHOW)	(*a*) For sharing the latest research findings with the field workers (*b*) Gives opportunity to dairy practitioners to present and discuss field problems	Once a year	Regular	Very little

(Contd...)

Table Contd...

Sl. No.	Linkage Mechanisms	Purpose	Frequency of Activities	Status of Activities with	
				NDRI Personnel	SDAH Personnel
		(c) Develops an orientation toward research to solve field problems			
5.	Farming Systems Research Project (FSRA)/Operational Research Project (ORP)	(a) Demonstration and testing the field applicability of improved package of practice (b) Providing insight to the scientists about the problems involved in applicability of these technologies under field conditions	Round the year activities (Currently the project has been terminated)	Very Little	Nil
6.	Dairy Mela	Offers common plateform for researchers, field functionary and farmers to interact with each other	Depends on the decision of the Institute		
7.	Institute Village Linkage Programme (IVLP)	Technology assessment and refinement for the development of appropriate transferable technologies	Round the year activities	Regular	Nil

8.1.2 Structural Linkage Mechanism Between Research and Extension as Developed by HAU

From the information presented in Table 8.2, it could be observed that HAU was marginally ahead of NDRI with respect to the number of SLM for RP and EP. The frequency of activities of these SLM showed almost similar pattern as that of NDRI, however, status of the activities of most of these SLM was regular with the personnel from HAU as well as SDAH. From the findings, it could be inferred that RP from HAU were structurally better linked with the EP from HAU as well as SDAH as compared to RP from NDRI. A relatively better status of SLM in HAU would definitely have influence on the strength of functional linkages between them (Table 8.2).

8.1.3 Structural Linkage Mechanism Between Research and Extension as Developed by SDAH

On investigation it was found that SDAH had only three SLM in order to interact with the RP. With the RP from NDRI, SDAH had nil and negligible activities during district level regional meeting and fertility camp, respectively. However, SDAH regularly invited the RP from HAU in fertility camp and *pasu mela* (Table 8.3).

From the above findings, it could be inferred that both HAU and SDAH had outperformed NDRI with respect to the number of SLM, and their status of activities. The probable reasons behind such difference could be the mandate of the respective organizations. Moreover, animal husbandry and dairying being the state's subject, better status of SLM between RP and EP by HAU and SDAH is comprehensible.

8.1.4 Structural Linkage Mechanism Between Extension and Farmers as Developed by the Selected Departments

Some of the SLM as developed by the selected extension systems were identified and the frequency and status of their activities were ascertained from the respective extension managers. The findings are briefly contained in Table 8.4.

Table 8.2. : Structural linkage mechanism between research and extension as developed by HAU

Sl. No.	*Linkage Mechanisms*	*Purpose*	*Frequency of Activities*	*Status of Actual Activities with*	
				HAU Personnel	*SDAH Personnel*
1.	Research Review Committee (RRC)	(*a*) Discuss the progress of research project (*b*) Offers farmers feedback to the researchers (*c*) Design and implement the multi-disciplinary research project	Twice a year	Regular	Regular
2.	Extension Advisory Committee (EAC)	(*a*) Decisions regarding planning and execution of extension activities (*b*) Monitoring and evaluation of the on-going and completed extension activities	Twice a year	Regular	Regular
3.	Extension Committee (EC)	Deals with those needs of farmers that require immediate attention and action	Need based	As and when such needs arise	As and when such needs arise
4.	Training Advisory Committee (TAC)	Imparts the field personnel with the technical know how	Annual	Not regular	Not regular

(Contd...)

Sl. No.	Linkage Mechanisms	Purpose	Frequency of Activities	Status of Actual Activities with	
				HAU Personnel	SDAH Personnel
5.	Clinical Conference	Exposes field personnel to the advances of clinical know how	Annual	Regular	Regular
6.	Joint Diagnostic Team	Studies and identifies the field problems jointly	Not fixed	Not regular	Very little
7.	Field Days	(*a*) Transfer of technology (*b*) Provides opportunity for interaction	Occasionally	Very little	Very little
8.	Institute Village Linkage Programme	Technology assessment and refinement for the development of appropriate transferable technology	Round the year activities	Regular	Almost nil
9.	Pasu Mela	Offers informal platform for researchers, field personnel and farmers	Once a year	Regular	Regular

Table 8.3 : Structural linkage mechanism between research and extension as developed by SDAH

Sl. No.	*Linkage Mechanisms*	*Purpose*	*Frequency of Activities*	*Status of Actual Activities with*	
				NDRI Personnel	*HAU Personnel*
1.	Regional Meeting at District Level	(*a*) Discuss the progress of field level activities (*b*) Discussion on the field level problems	Twice a year	Negligible	Very little
2.	Fertility Camp	(*a*) Diagnosis and treatment of infertility cases (*b*) Education the farmers about animal health management	Once-Twice a month	Nil	Regular
3.	District Pasu Mela and Cattle Show	Provides informal opportunity of interaction for extension personnel, researchers and farmers	Once a year	Regular	Regular

From the table, it could be observed that both NDRI and SDAH were regular with regard to fertility camp and veterinary aid camp. Further, NDRI was also found regular with respect to training about improved dairy management to the farmers and demonstration of practices to the dairy farmers for speedy adoption. HAU, on the other hand, was regular in the activities of SLM like group discussion/meetings with dairy farmers, training, field days, demonstration and field trials. On all the remaining SLM, status of activities by SDAH was observed rare to sometimes.

From the findings, it could be noted that SDAH had better status of activities on those SLM which could offer only technical inputs and services to the dairy farmers. Similarly, the component of supply and services was found absent in the SLM as developed by HAU and it emphasized mostly on the educational component. With respect to NDRI, however, it could be noted that it had relatively better status of activities of SLM which embodied both education to the farmers as well as ensuring technical inputs and services to them.

8.2 Extent of Functional Linkage Between Research, Extension and Dairy Farmers

Investigating into the extent of functional linkage (EOFL) between and among the entities, *viz.*, research personnel, extension personnel and dairy farmers was the cardinal pillar of the study. The functional linkage was measured by using the index developed for the same and accordingly the EOFL was computed. Under the following subheads, findings have been presented and discussed.

8.2.1 Extent of Functional Linkage Between Research and Extension Personnel

The EOFL between research and extension personnel was studied under two organizational systems, *viz.*, when both were in same organization (system one) and both operated from separate organisations (system two). Based on the overall functional linkage between them, they were distributed

Table 8.4. : Structural linkage mechanism between extension and farmers as developed by the selected department/organizations and their status activities

Sl. No.	*Linkage Forum/Activity*	*Status of Activities by*		
		NDRI	*HAU*	*SDAH*
1.	Fertility Camp and Veterinary-Aid Camp	Regular	Sometimes	Regular
2.	Group Discussion/Meetings with Dairy Farmers	Sometimes	Regular	Sometimes
3.	Training about Improved Dairy Management	Regular	Regular	Sometimes/Rare
4.	Field Days	Sometimes	Regular	Rare
5.	Exhibition/Shows/ Competition	Sometimes	Rare	Rare
6.	Demonstration	Always/Regular	Always/Regular	Rare
7.	Field/Adaptive Trials	Sometimes	Always/Regular	Never
8.	Study/Survey	Sometimes	Sometimes	Rare
9.	Radio/TV Talk	Rare	Sometimes	Rare
10.	Mela	Sometimes	Sometimes	Sometimes

according to their varying strength of the same. The findings so obtained are presented in Table 8.5. Findings in the table indicate that most of the respondents (63.49%) were found to be in "weak linkage" followed by 26 per cent in "moderate linkage" and little less, *i.e.*, about 25 per cent in "no linkage" category in the organization system, where both were operating together. However, in another setup of organization, equal percentages of the respondents (40.62%) were falling in "no linkage" and "weak linkage" category. Under this system, about 11 and 8 per cent of the respondents were observed to have "moderate linkage" and "strong linkage", respectively. From the same table, it is further clear that HAU outperformed NDRI in the strength of linkage in both the cases, *i.e.*, within the organization and with the SDAH. It could be hence inferred from the table, the strength of functional linkage would be better, if both research and extension are operating from the same organization. However, if a good structural mechanism is developed, the strength could be expected better, even if both are functioning from the separate organizations, as evident from the same table. The findings get support from the works of Bourgeosis (1989), Kessaba (1989), Pineiro (1989) and Antholt (1990). However, it is in contradiction with the work by Trents (1989) and Eponou (1993). The overall functional linkage between sampled research and extension personnel from the selected organization system were further broken into its various functional components. The findings have been presented in Tables 8.6 to 8.10 and discussed thereafter.

8.2.1.1 Communication Linkage Between Research and Extension Personnel

From Table 8.6, it is evident that a "strong communication linkage" was perceived by the 19 and 14 per cent of the selected research and extension personnel under the system one and system two, respectively. There were good percentage of respondents, *i.e.*, about 37 and 31 per cent who expressed "moderate strength" of functional linkage between them under the selected systems, respectively. From the same table, it is further revealed that highest percentage of respondents

Table : 8.5 : Frequency distribution of research and extension personnel selected from different organizational systems on the basis of their extent of overall functional linkage

Selected Organizational Systems	*Selected Organization/ Department and Personnel*	*Extent of Linkage*			
		No Linkage (0)	*Weak Linkage (<4.42)*	*Moderate Linkage (4.42-22.12)*	*Strong Linkage (>22.12)*
Research and extension in same organization	(*i*) Research Personnel				
	(*a*) NDRI (N=17)	5(29.41)	8(47.06)	4(23.53)	—
	(*b*) HAU (n=15)	2(13.33)	8(53.34)	5(33.33)	—
	Pooled (n=32)	7(21.88)	16(50.00)	9(28.12)	
	(*i*) Extension Personnel :				
	(*a*) NDRI (N=16)	6(37.50)	6(37.50)	4(25.00)	—
	(*b*) HAU (n=15)	3(20.00)	8(53.33)	4(26.67)	—
	Pooled (n=31)	9(29.03)	14(45.16)	8(25.81)	-
	Overall (n=63)	16(25.40)	40(63.40)	17(26.01)	-
Research and extension in separate organization	(*i*) Research Personnel				
	(*a*) NDRI with SDAH (n=17)	12(70.59)	5(29.41)		
	(*b*) HAU with SDAH (n=15)	4(26.67)	6(40.00)	3(20.00)	2(13.33)
	Pooled (n=32)	16(50.00)	11(34.38)	3(9.37)	2(6.25)
	(*i*) Extension Personnel :				
	(*a*) SDAH with NDRI (n=16)	6(37.50)	10(62.50)		
	(*b*) SDAH with HAU (n=16)	4(25.00)	5(31.25)	4(25.00)	3(18.75)
	Pooled (n=32)	**10(31.25)**	**15(46.87)**	**4(12.50)**	**3(9.38)**
	Overall (n=64)	**26(40.62)**	**26(40.62)**	**7(10.94)**	**5(7.82)**

* *Figures in parentheses indicate percentage*

Table : 8.6 : Frequency distribution of research and extension personnel selected from different organizational systems on the basis of their extent of communication linkage.

Selected Organizational Systems	*Selected Organization/ Department and Personnel*	*Extent of Linkage*			
		No Linkage (0)	*Weak Linkage (<4.42)*	*Moderate Linkage (4.42-22.12)*	*Strong Linkage (>22.12)*
Research and extension in same organization	(*i*) Research Personnel				
	(*a*) NDRI (N=17)	5(23.53)	6(35.29)	5(29.41)	2(11.77)
	(*b*) HAU (n=15)	—	2(13.33)	7(46.67)	6(40.00)
	Pooled (n=32)	4(12.50)	8(25.00)	12(37.50)	8(25.00)
	(*i*) Extension Personnel :				
	(*a*) NDRI (N=16)	7(43.75)	5(31.25)	3(18.75)	1(6.25)
	(*b*) HAU (n=15)	—	4(26.67)	8(53.33)	3(20.00)
	Pooled (n=31)	7(22.58)	9(29.03)	11(35.48)	4(12.91)
	Overall (n=63)	11(17.46)	17(26.98)	23(36.50)	12(19.04)
Research and extension in separate organization	(*i*) Research Personnel				
	(*a*) NDRI with SDAH (n=17)	12(70.59)	3(17.65)	2(11.76)	3(20.00)
	(*b*) HAU with SDAH (n=15)	4(33.33)	2(13.33)	5(33.34)	
	Pooled (n=32)	17(53.13)	5(15.63)	7(21.87)	3(9.37)
	(*i*) Extension Personnel :				
	(*a*) SDAH with NDRI (n=16)	5(31.25)	6(37.50)	5(31.25)	6(37.50)
	(*b*) SDAH with HAU (n=16)	2(12.50)		8(50.00)	
	Pooled (n=32)	**7(21.87)**	**6(18.75)**	**13(20.63)**	**6(18.75)**
	Overall (n=64)	**24(37.50)**	**11(17.19)**	**20(31.25)**	**9(14.06)**

* *Figures in parentheses indicate percentage*

Table : 8.7 : Frequency distribution of research and extension personnel selected from different organizational systems on the basis of their extent of linkage in collaborative professional activities

Selected Organizational Systems	*Selected Organization/ Department and Personnel*	*Extent of Linkage*			
		No Linkage (0)	*Weak Linkage (<4.42)*	*Moderate Linkage (4.42-22.12)*	*Strong Linkage (>22.12)*
Research and extension in same organization	(*i*) Research Personnel				
	(*a*) NDRI (N=17)	5(23.53)	6(35.29)	6(35.29)	—
	(*b*) HAU (n=15)	4(26.67)	3(20.00)	8(53.33)	—
	Pooled (n=32)	9(28.12)	9(28.12)	14(43.76)	—
	(*i*) Extension Personnel :				
	(*a*) NDRI (N=16)	6(37.50)	4(25.00)	6(37.50)	2(13.33)
	(*b*) HAU (n=15)	3(20.00)	3(20.00)	7(46.67)	—
	Pooled (n=31)	9(29.03)	7(22.58)	13(41.94)	2(6.45)
	Overall (n=63)	18(28.57)	16(25.39)	27(42.86)	2(3.18)
Research and extension in separate organization	(*i*) Research Personnel				
	(*a*) NDRI with SDAH (n=17)	17(100.00)	—	6(40.00)	—
	(*b*) HAU with SDAH (n=15)	9(60.00)	—		
	Pooled (n=32)	26(81.25)	—	6(8.75)	—
	(*i*) Extension Personnel :				
	(*a*) SDAH with NDRI (n=16)	15(93.75)	1(6.25)	7(43.75)	1(6.25)
	(*b*) SDAH with HAU (n=16)	6(37.50)	2(12.50)		
	Pooled (n=32)	**21(65.23)**	**3(9.37)**	**7(21.87)**	**1(3.13)**
	Overall (n=64)	**47(73.44)**	**3(4.69)**	**13(20.31)**	**1(1.56)**

* *Figures in parentheses indicate percentage*

Table : 8.8 : Frequency distribution of research and extension personnel selected from different organizational systems on the basis of their extent of linkage in planning of research and extension activities

Selected Organizational Systems	*Selected Organization/ Department and Personnel*	*Extent of Linkage*			
		No Linkage (0)	*Weak Linkage (<4.42)*	*Moderate Linkage (4.42-22.12)*	*Strong Linkage (>22.12)*
Research and extension in same organization	(*i*) Research Personnel				
	(*a*) NDRI (N=17)	6(35.29)	5(29.42)	6(35.29)	1(6.67)
	(*b*) HAU (n=15)	8(53.33)	3(20.00)	3(20.00)	
	Pooled (n=32)	14(43.75)	8(25.00)	9(28.12)	1(3.13)
	(*i*) Extension Personnel :				
	(*a*) NDRI (N=16)	10(62.50)	5(31.25)	1(6.25)	2(13.33)
	(*b*) HAU (n=15)	4(26.67)	5(33.33)	4(26.67)	
	Pooled (n=31)	14(45.16)	10(32.26)	5(16.13)	2(6.45)
	Overall (n=63)	28(44.44)	18(28.58)	14(22.22)	3(4.76)
Research and extension in separate organization	(*i*) Research Personnel				
	(*a*) NDRI with SDAH (n=17)	17(100.00)	2(13.33)	—	—
	(*b*) HAU with SDAH (n=15)	13(86.67)		—	—
	Pooled (n=32)	30(93.75)	2(6.25)	—	—
	(*i*) Extension Personnel :				
	(*a*) SDAH with NDRI (n=16)	16(100.00)	—	7(43.75)	—
	(*b*) SDAH with HAU (n=16)	9(56.25)	—		—
	Pooled (n=32)	**25(78.12)**	—	**7(21.88)**	—
	Overall (n=64)	**55(85.93)**	**2(3.13)**	**7(10.94)**	—

* *Figures in parentheses indicate percentage*

expressed the communication linkage as "weak to absent". The communication linkage within HAU, and between HAU and SDAH was expressed comparatively better by the large number of personnel than that of same within NDRI, and between NDRI and SDAH. From, the findings, it is obvious that the organization system, where both research and extension were together, performed relatively better than the same in separate organizations. As more structural linkage mechanisms were under operation in former case, communication linkage was better. Still, the mean communication linkage was to the extent of only 30 per cent in case of former and 13 per cent in case of system two (Table 8.12) clearly indicate poor exploitation of the existing structural mechanism in case of system one and establishment of more mechanisms and better performance of existing mechanisms, hence in system two is warranted.

8.2.1.2 *Collaborative Professional Activities Between Research and Extension Personnel*

The second dimension of functional linkage between research and extension personnel studied was the collaborative professional activity. The findings contained in Table 8.7 indicate that most of the respondents (about 42%) in case of system one were under "moderate strength" of functional linkage, followed by "absent" and "weak linkage". However, in system two, as high as 73 per cent of the respondents felt "absent linkage" between them. In both the systems, HAU outperformed NDRI on this parameter of functional linkage between research and extension.

8.2.1.3 *Linkage in Planning and Decision-Making*

The third dimension of functional linkage, *i.e.*, planning and decision-making has been reported in Table 8.8. The perusal of table indicate as high as 44.44 and 78 per cent of the respondents expressed it as "absent", *i.e.*, no linkage and "weak" linkage, respectively. However, about 22 per cent and only 5 per cent of them observed the same as "moderate" and "strong", respectively.

Table : 8.9 : Frequency distribution of research and extension personnel selected from different organizational systems on the basis of their extent of linkage in implementation and evaluation of research and extension activities

Selected Organizational Systems	*Selected Organization/ Department and Personnel*	*Extent of Linkage*			
		No Linkage (0)	*Weak Linkage (<4.42)*	*Moderate Linkage (4.42-22.12)*	*Strong Linkage (>22.12)*
Research and extension in same organization	(*i*) Research Personnel				
	(*a*) NDRI (N=17)	10(58.82)	4(23.53)	3(17.65)	—
	(*b*) HAU (n=15)	10(66.67)	2(13.33)	3(20.00)	—
	Pooled (n=32)	20(62.50)	6(18.75)	6(18.75)	—
	(*i*) Extension Personnel :				
	(*a*) NDRI (N=16)	10(62.50)	4(25.00)	2(12.50)	—
	(*b*) HAU (n=15)	5(33.33)	4(26.67)	6(40.00)	
	Pooled (n=31)	15(43.40)	8(25.81)	8(25.81)	—
	Overall (n=63)	35(55.56)	14(22.22)	14(22.22)	—
Research and extension in separate organization	(*i*) Research Personnel				
	(*a*) NDRI with SDAH (n=17)	17(100.00)	—	2(13.33)	—
	(*b*) HAU with SDAH (n=15)	13(86.67)		—	—
	Pooled (n=32)	30(93.75)	—	2(6.25)	—
	(*i*) Extension Personnel :				
	(*a*) SDAH with NDRI (n=16)	16(100.00)	—	6(37.59)	—
	(*b*) SDAH with HAU (n=16)	10(62.50)	—		—
	Pooled (n=32)	**26(81.25)**	—	**6(18.75)**	—
	Overall (n=64)	**56(87.50)**	—	**8(12.50)**	—

* *Figures in parentheses indicate percentage*

Table : 8.10 : Frequency distribution of research and extension personnel selected from different organizational systems on the basis of their extent of linkage in training

Selected Organizational Systems	*Selected Organization/ Department and Personnel*	*Extent of Linkage*			
		No Linkage (0)	*Weak Linkage (<4.42)*	*Moderate Linkage (4.42-22.12)*	*Strong Linkage (>22.12)*
Research and extension in same organization	(*i*) Research Personnel				
	(*a*) NDRI (N=17)	10(58.82)	7(41.18)	—	—
	(*b*) HAU (n=15)	10(66.67)	2(13.33)	3(20.00)	—
	Pooled (n=32)	20(62.50)	9(28.12)	3(9.38)	—
	(*i*) Extension Personnel :				
	(*a*) NDRI (N=16)	13(81.25)	3(18.75)	—	—
	(*b*) HAU (n=15)	6(40.00)	4(26.67)	5(33.33)	
	Pooled (n=31)	19(61.29)	7(25.58)	5(16.13)	—
	Overall (n=63)	39(61.90)	16(25.40)	8(12.70)	—
Research and extension in separate organization	(*i*) Research Personnel				
	(*a*) NDRI with SDAH (n=17)	17(100.00)			—
	(*b*) HAU with SDAH (n=15)	9(60.00)	3(20.00)	3(20.00)	—
	Pooled (n=32)	26(81.26)	3(9.37)	3(9.37)	—
	(*i*) Extension Personnel :				
	(*a*) SDAH with NDRI (n=16)	16(100.00)			—
	(*b*) SDAH with HAU (n=16)	8(50.00)	5(31.25)	3(18.75)	—
	Pooled (n=32)	**24(75.00)**	**5(15.62)**	**3(9.38)**	—
	Overall (n=64)	**50(78.12)**	**8(12.50)**	**6(9.38)**	—

* *Figures in parentheses indicate percentage*

8.2.1.4 *Linkage in Joint Implementation and Evaluation*

As the status of linkage in joint planning and decision making was discouraging, the strength of same in joint implementation and evaluation of any research and extension activity was in the similar line. The data pertaining to this aspect has been presented in Table 8.9. The findings revealed that though the high percentage (about 78%) of the respondents categorically found the strength of linkage as "weak" and "no linkage"; 22 per cent of them were falling in the "moderate" strength category. The scenario of system two was even more discouraging and as high as 88 per cent of the respondents found that the linkage was "absent" between them. A relatively better performance of structural mechanism between HAU and SDAH was the reason that 12.50 per cent of the respondents of system two found "moderate" strength of linkage between them on the said parameter.

8.2.1.5 *Linkage in Training*

The fifth aspect of functional linkage between research and extension personnel, *i.e.*, linkage in training was studied in the similar style. The perusal of Table 8.10 indicates that a "moderate" linkage was expressed by only about 13 and 9 per cent of the respondents operating under system one and system two, respectively. A large percentage (as high as 78%) of them reported "absent" linkage between them as far as training was concerned. Within HAU, however, relatively more number of respondents (33.33%) perceived a "moderate" strength of linkage compares to between HAU and SDAH (20%). The NDRI displayed disheartening picture on this parameter, both within the organization as well as with SDAH.

8.2.1.6 *Linkage in Supply and Services*

As was the case with functional linkage in training, the linkage between research and extension personnel with respect to supply and service was quite discouraging. As a result, not more than 19 per cent of the respondents, that too from system one, could express a "moderate" strength of functional linkage between them. Most of them observed it to be "absent to weak". All the functional activities of training and supply services were being done from rare to never.

From the findings explained in the above paragraphs, it could be observed that almost all parameters of functional linkage between research and extension personnel were non functional to a larger extent. Among those, only communication linkage and collaborative professional activities were observed to be marginally better performed. The other aspects like planning and decision making, implementation and evaluation, training, and supply services did exceptionally poor. This leads to inference that the existing level of structural mechanisms are not adequate with respect to their mandate and mode of participation of research and extension personnel in them. Very poor strength of functional linkage between research and extension in animal husbandry was also reported by Singh (1994) and Gupta (1998) and they also attributed this to the non existence of proper mechanism and poor performance of existing structural mechanism in the organization.

From the findings, it could also be noted that though the overall functional linkage as well as the same on various parameters was "absent to weak" or in some cases "moderate", the organization system where both research and extension were operating together outperformed the system, where research and extension were in separate organizations. This heavily comes down upon the existing linkage mechanisms between the organizations/department under system two. Though from Tables 8.1 to 8.3, it could be observed that a good number of structural mechanisms were under existence, the very poor strength of functional linkage indicate either the non performance by them or conforming to the age old organizational mandates/goals.

The strength of functional linkage between research and extension was worked out in percentage and presented in Table 8.11. From the table, it is evident that MEFL was only 13.27 in organization system, where both the research and extension were together. A high percentage of coefficient of variation (CV; 85.15%) indicates the greater variation in the participation of

Table : 8.11 : Mean of overall extent of functional linkage (%) between research and extension personnel under selected organizational systems

S. No.	*Selected Organizational Systems*	*Selected Organization/ Department*	*Mean of Overall Extent (%) of Functional Linkage*	*S.D.*	*C.V. (%)*
1.	Research and extension in same organization	(*i*) Within NDRI (n=33)	12.19	9.07	74.40
		(*ii*) Within HAU (n=30)	16.10	18.74	54.28
		Overall (n=63)	13.27	11.30	58.15
2.	Research and extension in separate organization	(*i*) Between NDRI and SDAH (n=33)	1.16	0.85	73.27
		(*ii*) Between HAU and SDAH (n=31)	10.19	4.25	41.90
		Overall (n=64)	5.58	10.16	172.50

research and extension personnel in various linkage related activities. Under this system, HAU outperformed the NDRI as the MEFL within former was 16.10 and 12.19 per cent in later. Relatively lower CV further revealed that the personnel of HAU varied comparatively lower than those of NDRI with respect to their participation in linkage activities. The scenario of MEFL under the system two was quite discouraging. The overall MEFL was computed to only 5.89 per cent with 172.50 per cent of CV. The MEFL between HAU and SDAH was to the extent of 10.19 per cent and the same between NDRI and SDAH was only 1.16 per cent. The CV was found to the extent of 73.27 per cent in the later case. In order to give statistical validation of the difference of MEFL between the selected organization systems, the MEFLs of selected indicators were subjected to unpaired *t* test. The findings, as contained in Table 8.12 says that barring only supply and services, the MEFL on all the parameters was significantly more in the system one than the system two.

Table 8.12 : Comparison of the mean of functional linkages between research and extension in the selected organizational systems.

S.No.	*Parameters of Functional Linkage*	*Research and Extension in same Organization (n_1=63)*		*Research and Extension in Separate Organization (n_2=64)*		$\lvert \overline{X}_1 - \overline{X}_2 \rvert$	*t-Value at 125 d.f.*
		Mean ($\overline{X}_1$)	*S.D.*	*Mean* ($\overline{X}_2$)	*S.D.*		
1.	Communication	30.29	19.07	12.86	15.01	17.43	12.78*
2.	Collaborative professional activities	17.25	25.10	6.82	13.42	10.43	13.37*
3.	Planning and decision making	9.24	11.71	2.33	6.62	6.91	11.73*
4.	Implementation and evaluation	8.07	10.57	2.34	8.32	5.73	10.52*
5.	Training	11.27	12.52	4.33	10.40	6.94	11.55*
6.	Supply and services	5.49	7.17	4.56	10.49	0.93	1.76^{NS}
7.	Overall	13.27	11.30	5.89	10.16	7.38	12.69*

* = Significant at 1 per cent level of significance NS = Not significant

Further, under each system, different organisations/ departments were selected. Hence, it was imperative to compare the MEFL between HAU and NDRI which were the two constituents of system one; and between HAU with SDAH and NDRI with SDAH which constituted system two. Unpaired t-statistics was employed to see the significance of difference. From Table 8.13, it is clear that except collaborative professional activities, the MEFL was significantly more in case of HAU than NDRI. The findings lead to inference, that HAU had better edge over NDRI, while later performed a comparable level on collaborative professional activities. This further leads to

Table 8.13 : Comparison of the means of functional linkages between research and extension in the organizational systems where both are operating together.

S.No.	Parameters of Functional Linkage	NDRI with its own extension system (n_1=33)		HAU with its own extension system (n_1=30)		$\lvert \overline{X}_1 - \overline{X}_2 \rvert$	t-Value at 61 d.f.
		Mean(X_1)	S.D.	Mean(X_2)	S.D.		
1	2	3	4	5	6	7	8
1.	Communication	25.15	15.82	34.84	18.83	9.69	9.21*
2.	Collaborative professional activities	17.87	30.71	17.85	19.12	0.02	0.02^{NS}
3.	Planning and decision making	6.87	9.79	11.78	13.18	4.91	5.72*
4.	Implementation and evaluation	6.61	9.96	9.96	10.57	3.35	4.14*
5.	Training	11.80	11.95	10.74	12.52	1.06	12.00*
6.	Supply and services	2.01	3.52	9.47	10.79	7.46	10.93*
7.	Overall	10.07	8.83	16.75	12.13	6.68	8.15*

* = Significant at 1 per cent level of significance NS = Not significant

inference that though HAU has more number of structural linkage mechanisms between research and extension than NDRI, almost similar level of MEFL between them indicate a dissatisfactory performance of these mechanisms in HAU. Under the system two, comparison was made between two of its sub components. From Table 8.14, it is evident that on all the dimensions of functional linkage, the MEFL of HAU with SDAH was significantly more than that of NDRI with SDAH. The reasons were obvious. Firstly, agriculture and dairying being state subjects, HAU and SDAH have developed a good number of structural mechanisms than NDRI. Secondly, the mandate of NDRI might not be permitting her to make frequent interaction with SDAH on the selected parameters of the functional linkage between them.

Table 8.14 : Comparison of the means of functional linkages between research and extension in the organizational systems where both are operating separately

S. No.	*Parameters of Functional Linkage*	*NDRI with SDAH (n_1=33)*		*HAU with SDAH (n_2=31)*		$\lvert\overline{X}_1 - \overline{X}_2\rvert$	*t-Value at 61 d.f.*
		Mean ($\overline{X}_1$)	*S.D.*	*Mean* ($\overline{X}_2$)	*S.D.*		
1.	Communication	5.39	7.79	21.48	17.59	16.09	17.95*
2.	Collaborative professional activities	0.20	1.16	13.44	16.55	13.24	17.55*
3.	Planning and decision making	0.40	1.52	4.27	8.94	3.87	6.69*
4.	Implementation and evaluation	0.00	0.00	4.84	11.54	484	13.00*
5.	Training	0.50	2.02	8.87	13.76	8.37	11.78*
6.	Supply and services	0.00	0.00	8.96	13.58	8.96	20.46*
7.	Overall	1.17	1.85	10.92	12.73	9.75	14.27*

* = Significant at 1 per cent level of significance

8.2.2 Extent of Functional Linkage Between Extension Personnel and Dairy Farmers

The second important part of functional linkage among the three selected actors of development was studied in terms of reciprocal interaction between extension personnel (EP) and the dairy farmers (DFs). For more comprehensive investigation, three types of extension systems, *viz.*, HAU, NDRI and SDAH were selected and their respective clients were chosen for this purpose. Response on the common instruments was elicited by them. Based on the obtained response, mean extent of functional linkage (MEFL) was computed and their categorization was done separately. Findings have been presented and discussed indicator-wise under the following subheads :

8.2.2.1 *Communication Linkage Between Extension Personnel and Dairy Farmers*

From Table 8.15, it is evident that as high as 60 per cent and not less than 50 per cent of the selected EP had expressed their "moderate" strength of communication linkage with the DFs. Fifty per cent of the EP from SDAH expressed "strong" linkage with their clients. However, only 31 and 25 per cent from NDRI and HAU, respectively, had a sense of "strong" linkage with the farmers.

It could also be observed that about 13 and 20 per cent of the sampled EP from NDRI and HAU, respectively, perceived "weak" strength of communication linkage with the farmers. The average extent of communication linkage (AECL) of EP with DFs were computed about 50, 44 and 58 per cent in the NDRI, HAU and SDAH extension system, respectively. However, this was found to be 50.44 per cent on an overall basis (Table 8.16). When the MEFL were subjected to least squares ANOVA, variation of the same in the three selected systems was found non significant (Table 8.17). This revealed that except communication linkage, all the three selected extension systems displayed a similar level of performance on other parameters of functional linkage with their clients.

In order to counter validate the response of EP, average extent of communication linkage (AEC) for dairy farmers (DFs) was computed and they were distributed based on their differential AECL with EP.

Table 8.15 : Frequency distribution of extension personnel sampled from the selected extension systems based on their extent of linkage with farmers

Extension Systems	*Extent of Functional*	*Communication*	*Planning & Decision Making*	*Implementation and Evaluation*	*Supply & Service*	*Training*	*Overall Linkage*
NDRI (n=16)	N (O)	0(0.00)	1(6.25)	0(0.00)	0(0.00)	3(18.75)	0(0.00)
	W(<28.75)	2(12.50)	5(31.25)	2(12.50)	0(0.00)	8(50.00)	0(0.00)
	M(28.75-60.45)	9(56.25)	7(43.75)	11(68.75)	3(18.75)	3(18.75)	13(18.25)
	S(>60.45)	5(31.25)	3(18.75)	3(18.75)	13(81.25)	2(12.50)	3(18.75)
HAU (n=15)	N (O)	0(0.00)	1(6.67)	1(6.67)	0(0.00)	0(0.00)	0(0.00)
	W (<28.75)	3(20.00)	4(26.67)	2(13.33)	6(40.00)	10(66.67)	4(26.67)
	M (28.75–60.45)	9(60.00)	8(53.33)	10(66.67)	7(46.67)	3(20.00)	11(73.33)
	S (>60.45)	3(20.00)	2(13.33)	2(13.33)	2(13.33)	2(13.33)	0(0.00)
SDAH (n=16)	N (O)	0(0.00)	3(18.75)	2(12.50)	0(0.00)	3(18.75)	0(0.00)
	W (<28.75)	0(0.00)	2(12.50)	4(25.00)	0(0.00)	6(37.50)	3(18.75)
	M (28.75–60.45)	8(50.00)	9(56.25)	9(56.35)	2(12.50)	5(31.25)	10(62.50)
	S (>60.45)	8(50.00)	2(12.50)	1(6.25)	14(87.50)	2(12.50)	3(18.75)
Pooled (n=47)	N (O)	0(0.00)	5(10.65)	3(6.38)	0(0.00)	6(12.77)	0(0.00)
	W (<28.75)	5(10.64)	11(23.40)	8(17.02)	6(12.77)	24(51.06)	7(14.80)
	M (28.75–60.45)	26(55.32)	24(51.06)	30(63.83)	12(25.53)	11(23.40)	34(72.34)
	S (>60.45)	16(35.04)	7(14.89)	6(12.77)	29(12.77)	6(12.77)	6(12.77)

N = No : W = Weak : M = Moderate : S = Strong

Figures in parentheses indicate percentage.

Table 8.16 : Average extent of functional linkage (%) of extension personnel of the selected extension systems with the dairy farmers on various parameters

S.No.	Extension Personnel Taken from	Communication	Planning & Decisions making	Implementation & Evaluation	Supply and Services	Training	Overall
1.	NDRI (n=16)	49.65	38.02	46.34	80.72	20.11	47.48
2.	HAU (n=15)	43.70	38.89	42.85	38.43	38.43	38.81
3.	SDAH (n=16)	57.97	32.82	38.10	79.17	28.54	45.25
4.	Overall/ Pooled (n=47)	50.44	33.57	42.43	66.11	29.06	43.85

Table 8.17 : Least–squares ANOVA for the means of functional linkages of extension personnel with farmers of the selected extension systems

S.No.	Source	d.f.	S.S.	M.S.S.	F-Value
1.	Communication	2	1022.5676	511.2838	2.3180 NS
2.	Planning and decision making		33.0106	16.5053	0.4158 NS
3.	Implementation and evaluation	2	303.1673	151.5836	0.5892 NS
4.	Supply and services	2	17626.6971	8813.3486	28.7545**
5.	Training	2	617.9884	308.9942	1.0438 NS
6.	Overall	2	733.6309	366.81546	1.4897NS

** = Significant at 5 per cent level of significance. NS = Non-significant

From Table 8.18, it could be observed that when the overall scenario was taken, about 49 per cent of DF were in "moderate" linkage, followed by 24 per cent in "weak" linkage, 17 per cent in "strong" linkage, and 10 per cent in "no" linkage category. It is also clear from the table that a good range of the respondents (45 12%) from HAU extension system expressed "no" linkage with their EP. In this system, little variation could be noted in the percentage of respondents having "weak" and "moderate" strength of communication linkage with the extension personnel. The overall AECL was found to be 25 per cent (Table 8.23). For the pooled data, AECL were 32.37, 11.51 and 29.77 per cent in case of the dairy farmers of NDRI, HAU and SDAH, respectively.

The variation in the AECL between the selected systems as well as between the category was found significant (at 5% level of significance; Table 8.24). It means that the dairy farmers had better communication linkage with their extension personnel in the adopted villages of NDRI and SDAH as compared to HAU.

Table 8.18 : Frequency distribution of dairy farmers sampled from the selected extension systems based on their extent of overall linkage with the extension personnel

Extension Systems	*Category of Dairy Farmers*	*Extent of Overall Linkage*			
		No Linkage (0)	*Weak Linkage (<11.89)*	*Moderate Linkage (11.89 to 31.11)*	*Strong Linkage (>31.11)*
NDRI	Small (n=21)	–	8(38.09)	11(52.38)	2(9.53)
	Medium (n=27)	–	2(7.41)	20(74.07)	5(18.52)
	Large (n=12)	–	0(0.00)	6(50.00)	6(50.00)
	Pooled (n=60)	**–**	**10(16.67)**	**37(61.67)**	**13(21.66)**
HAU	Small (n=22)	10(45.45)	7(31.82)	5(22.73)	0(0.00)
	Medium (n=25)	3(12.00)	11(44.00)	11(44.00)	0(0.00)
	Large (n=13)	5(38.46)	3(23.08)	3(23.08)	2(15.38)
	Pooled (n=60)	18(30.00)	21(35.00)	19(31.67)	2(3.33)
SDAH	Small (n=20)	–	5(25.00)	12(60.00)	3(15.00)
	Medium (n=25)	–	7(28.00)	10(40.00)	8(32.00)
	Large (n=15)	–	0(0.00)	10(66.67)	5(33.33)
	Pooled (n=60)	–	12(20.00)	32(53.33)	16(26.67)
Overall	Small (n=63)	10(15.87)	20(31.75)	28(44.44)	5(7.94)
	Medium (n=77)	3(3.90)	20(25.97)	41(53.25)	13(16.88)
	Large (n=40)	5(12.50)	3(7.50)	19(47.50)	13(32.50)
	Pooled (n=180)	18(10.00)	43(23.89)	88(48.89)	31(17.22)

Figures in parentheses indicate percentage.

However, the variation of AECL between the selected categories of the dairy farmers was found significant only in case of NDRI. In case of HAU and SDAH, this variation was noted non significant (Table 8.25). This lead to inference that the better off dairy farmers communicated well with the extension personnel of NDRI. However, SDAH paid attention to all dairy farmers irrespective of their category. When the mean score of communication linkage of EP (51.65) and DFs (26.82) were compared, a significant difference was observed

(Table 8.26). The percentage gap in the communication linkage between them computed to be about 48.07 which is quite high (Table 8.26).

From the above findings, it could be concluded that the extension systems having the village level institutions (as that of NDRI and SDAH) were better linked with their clients than the system lacking them (HAU). A considerable gap in the extent of communication linkage between the personnel and farmers is indicative of the fact that EP were communicating with a limited number of clients and the better offs were given priority. Eventhough, HAU has more number of structural mechanism to connect their EP with the farmers, non availability of village level institution could be the reason for relatively poor linkage strength.

8.2.2.2 *Linkage Between Extension Personnel and Dairy Farmers with Respect to Joint Planning and Decision Making*

Linkage in planning and decision making was operationalized as all the activities of extension to be planned and decided jointly by the EP and DFs. For the pooled data, findings in Table 8.15 reveal that most of the EP (51.06%) had "moderate" strength of linkage, followed by "weak" (23.40%), "strong" (14.89%) and "poor" (10.65%) extent of linkage with farmers in joint planning and decision making (JPDM). Systems wise also, almost similar trend was noted except in NDRI, where about 31 per cent of the EP perceived "weak" linkage. The overall MEFL in JPDM was 33.57 per cent with break ups of 38.02, 38.89 and 32.82 per cent in case of NDRI, HAU and SDAH, respectively (Table 8.16). From these figures, it appear that HAU had comparative edge over others on this parameter. Statistically, however, this variation was noted non significant (Table 8.17).

The findings discussed in the subhead of 8.15 reveals relatively poor strength of communication linkage of HAU personnel. But, it could be inferred out of the findings in Table 8.16 that despite the non provision of village level institution, HAU is striving sincerely to make people's participation in development activities. Still, a poor level of MEFL in JPDM clearly revealed the top down approach being followed by the selected extension systems.

Table 8.19 : Frequency distribution of dairy farmers sampled from the selected extension systems based on their extent of linkage in joint planning and decision making process with extension personnel

Extension Systems	*Category of Dairy Farmers*	*Extent of Linkage in joint planning and Decision Making*			
		No Linkage (0)	*Weak Linkage (<11.89)*	*Moderate Linkage (11.89 to 31.11)*	*Strong Linkage (>31.11)*
1	*2*	*3*	*4*	*5*	*6*
NDRI	Small (n=21)	14(66.67)	7(33.33)	–	–
	Medium (n=27)	15(55.56)	9(33.33)	3(11.11)	–
	Large (n=12)	4(33.33)	5(41.67)	3(25.00)	–
	Pooled (n=60)	33(55.00)	33(55.00)	6(10.00)	–
HAU	Small (n=22)	14(63.64)	5(22.73)	3(13.63)	–
	Medium (n=25)	7(28.00)	11(44.00)	4(16.00)	3(12.00)
	Large (n=13)	5(38.46)	4(30.77)	3(23.07)	1(7.70)
	Pooled (n=60)	26(43.33)	20(33.33)	10(16.67)	4(6.67)
SDAH	Small (n=20)	15(75.00)	4(20.00)	1(5.00)	–
	Medium (n=25)	9(36.00)	5(24.00)	8(32.00)	2(8.00)
	Large (n=15)	6(40.00)	5(33.33)	2(13.33)	2(13.33)
	Pooled (n=60)	30(50.00)	15(25.00)	11(18.33)	4(6.67)
Overall	Small (n=63)	43(68.25)	15(25.40)	4(6.35)	–
	Medium (n=77)	31(40.26)	26(33.77)	15(19.48)	5(6.49)
	Large (n=40)	15(37.50)	14(35.00)	8(20.00)	3(7.50)
	Pooled (n=180)	89(49.44)	56(31.11)	27(15.00)	8(4.45)

Figures in parentheses indicate percentages.

Farmers' perception of their strength of functional linkage with EP on JPDM was still poorer. As high as 49.44 and 31.11 per cent of the farmers expressed their "no linkage" and "weak linkage" with EP on this parameter. Only 15 per cent of them had "moderate" and as low as 4 per cent of them had "strong" linkage with the EP (Table 8.19). Systems wise, DFs of the SDAH displayed relatively better scenario. In all the systems, large farmers appeared to show comparatively good picture than the small and medium categories of farmers. The mean strength of functional linkage of the DFs in JPDM was computed to be as low as 5.75, 9.38 and 7.75 per cent in case of farmers of NDRI, HAU and SDAH, respectively. The overall values were 3.42, 7.02, 9.97 and 6.80 per cent for small, medium, large farmers and for pooled, respectively (Table 8.23). The variation of MEFL of DFs in JPDM was not significant either between the systems or between the categories. Even between the categories in the systems, this variation was not significant (Table 8.25).

The findings clearly reveals that irrespective of the status of the farmers, their linkage with EP in JPDM was greatly disheartening. When the mean functional linkage scores of EP and DFs were compared, highly significant difference was noted and the percentage gap in their differential perception was as high as 60.32 per cent (Table 8.26).

Preceding discussion though revealed the presence of good number of structural mechanisms between EP and DFs, findings of this subhead definitely lead to conclusion that these structural mechanisms were either product driven and top down or were non operational. In other words, most of activities of field extension works seem to be pre-decided and pre-planned irrespective of their acceptability by the farmers. Claim made by the EP about the involvement of DFs in JPDM appears to be rituals and mere formality.

8.2.2.3 *Linkage Between Extension Personnel and Dairy Farmers with Respect to Implementation and Evaluation*

This was measured in terms of extent of involvement of EP and DFs in the joint implementation and evaluation of field extension activities. Based on their response, MEFL in joint implementation and evaluation (JIE) was worked out and the respondents were categorized into different levels of strength of functional linkage.

From the findings presented in Table 8.15, it is clear that most of the EP (about 64%) were under "moderate" strength followed by 17.02, 13.00 and 6.00 per cent under "weak", "strong" and "no" linkage status, respectively, with the dairy farmers as far as joint implementation and evaluation of the extension activities was concerned. From the same table, it is also evident that the percentage of EP falling in "moderate" strength of functional linkage in JIE was marginally high in case of NDRI (68.75%) than HAU (66.67%) and SDAH (56.25%). Similarly, percentage of EP coming under "strong" linkage followed the similar trend. The average extent of functional linkage in JIE were 46.34, 42.85 and 38.10 per cent in case of NDRI, HAU and SDAH, respectively. The overall percentage was 42.43 per cent (Table 8.17).

It could be inferred from the findings discussed earlier that though EP had poor linkage with DFs in planning and deciding the extension programmes (33.57%), relatively better picture is revealed in case of their linkage with DFs in implementation of the field activities. All the three selected extension systems showed a consistent performance and hence, the variation of MEFL in JIE between, was found non significant.

Farmers' point of view about their extent of functional linkage with EP in JIE of extension activities showed a great variation. Majority of them (about 48%) were in "moderate" strength category, followed by 26.11, 14.44 and 11.67 per cent in "weak", "strong" and "no" linkage categories, respectively (Table 8.20).

On this parameter, majority of the farmers of the villages adopted by HAU were under "no linkage" and "weak linkage" category. Farmers of SDAH and NDRI were found to have better interaction with their EP in implementing and evaluating the extension activities jointly. The MEFL of the sampled farmers of NDRI, HAU and SDAH were 20.41, 14.95 and 25.39 per cent, respectively. Category wise small, medium and large dairy farmers were found to have 14.93, 21.45 and 24.42 per cent MEFL, respectively, with the EP (Table 8.23). The variation of MEFL between the categories of farmers and between the selected extension system was found statistically non significant (Table 8.24). However, the variation of the MEFL between the categories in NDRI system was found significant at 5 per cent level of significance, revealing due attention given on the large farmers during the implementation of field extension activity. When the overall MEFL, as perceived by EP (44.31%) and DF (11.62%) were subjected to unpaired *t* test, a highly significant difference was noted and also the percentage gap in linkage in JIE between EP and DFs was computed to the extent of about 50 per cent.

Table 8.20 : Frequency distribution of dairy farmers sampled from the selected extension systems based on their extent of linkage in Implementation and Evaluation with extension personnel

Extension Systems	*Category of Dairy Farmers*	*Extent of Linkage in Joint Planning and Decision Making*			
		No Linkage (0)	*Weak Linkage (<11.89)*	*Moderate Linkage (11.89 to 31.11)*	*Strong Linkage (>31.11)*
NDRI	Small (n=21)	2(9.52)	7(33.33)	11(57.15)	2(7.41)
	Medium (n=27)	–	8(29.63)	17(62.96)	2(16.66)
	Large (n=12)	–	2(16.67)	8(66.67)	4(6.67)
	Pooled (n=60)	2(3.33)	17(28.33)	37(61.67)	–
HAU	Small (n=22)	11(50.00)	4(18.18)	7(31.82)	3(12.00)
	Medium (n=25)	3(12.00)	9(36.00)	10(40.00)	1(7.69)
	Large (n=13)	5(38.46)	4(30.77)	3(23.08)	4(6.67)
	Pooled (n=60)	19(31.67)	17(28.33)	20(33.33)	–
SDAH	Small (n=20)	–	4(20.00)	12(60.00)	4(20.00)
	Medium (n=25)	–	7(28.33)	9(36.00)	9(36.00)
	Large (n=15)	–	2(31.32)	8(53.33)	5(33.34)
	Pooled (n=60)	–	30(21.67)	29(48.33)	18(30.00)
Overall	Small (n=63)	13(20.63)	15(23.81)	31(49.21)	4(6.35)
	Medium (n=77)	3(3.90)	24(31.17)	36(46.75)	14(18.18)
	Large (n=40)	5(12.50)	8(20.00)	19(47.50)	8(20.00)
	Pooled (n=180)	21(11.67)	47(26.11)	86(47.78)	26(14.44)

Figures in parentheses indicate percentages

8.2.2.4 *Linkage in Supply and Service*

Linkage between EP and DFs in supply and services (SS) was operationalized as the extent of their involvement in assessing the needs for technical inputs and services (TIPS), supply of the same and coöperation rendered during field supply of the technical inputs and services.

From the findings presented in Table 8.15, it is evident that as high as about 62 per cent of the EP were in the "strong" linkage category, followed by 25.53 and 12.77 per cent in the "moderate" and "weak" category. None of the EP felt absence of linkage with the DFs on this parameter. From the same table, it is also clear that as high as 87.50 and 81.25 per cent of the personnel from SDAH and NDRI were in "strong" linkage category. In case of HAU, this percentage was only 13.33 and majority of EP were in "moderate" (46.67%) and "weak" (40.00%) linkage strength. Consequently, the mean strength of functional linkage in SS was worked out to be 80.72, 79.17 and 38.43 per cent in case of NDRI, SDAH and HAU, respectively. The overall strength of the linkage on this aspect was 66.11 per cent (Table 8.16). Statistically, the difference of MEFL in SS of the selected extension systems was found to be significant at 5 per cent level of significance.

Farmers' perception of their strength of functional linkage in SS with EP was ascertained and based on their response, categorisation was done. The distribution of farmers coming under different strength category has been contained in Table 8.21.

The perusal of this table reveals that majority (58.89%) of them were in "strong" linkage irrespective of their categories, followed by "moderate" (21.67%), "no linkage" (10.56%) and "weak" (8.88%) linkage categories. The table also reveals that as high as 91.67 and 75.00 per cent of the beneficiaries from SDAH and NDRI, respectively, were in "strong" linkage category. However, majority of the farmers from the villages adopted by HAU were either in "no linkage" (31.67%) or "weak linkage" (26.66%) category. The mean strength of functional linkage of DFs in SS was found to be 57.41, 11.98 and 64.29 per cent for the dairy farmers of NDRI, HAU and SDAH, respectively. The overall value was 44.56 per cent. Category wise, the mean values were 37.06, 48.57 and 48.06 per cent for small, medium and large farmers, respectively (Table 8.23).

Table 8.21 : Frequency distribution of dairy farmers sampled from the selected extension systems based on their extent of linkage in Supply and Service with extension personnel

Extension Systems	*Category of Dairy Farmers*	*Extent of Linkage in Joint Planning and Decision Making*			
		No Linkage (0)	*Weak Linkage (<11.89)*	*Moderate Linkage (11.89 to 31.11)*	*Strong Linkage (>31.11)*
NDRI	Small (n=21)	–	–	6(28.57)	15(71.43)
	Medium (n=27)	–	–	7(25.93)	20(74.07)
	Large (n=12)	–	–	2(16.67)	10(89.33)
	Pooled (n=60)	–	–	15(25.00)	45(75.00)
HAU	Small (n=22)	11(50.00)	6(27.27)	3(13.64)	3(12.00)
	Medium (n=25)	3(12.00)	7(28.00)	13(52.00)	1(7.69)
	Large (n=13)	5(38.46)	3(23.07)	3(23.07)	2(6.00)
	Pooled (n=60)	19(31.67)	16(26.66)	19(31.67)	6(10.00)
SDAH	Small (n=20)	–	–	5(25.00)	15(75.00)
	Medium (n=25)	–	–	–	75(100.00)
	Large (n=15)	–	–	–	15(100.00)
	Pooled (n=60)	–	–	5(8.33)	55(91.67)
Overall	Small (n=63)	11(17.43)	6(9.53)	14(22.22)	4(6.35)
	Medium (n=77)	3(3.90)	7(9.10)	20(25.87)	47(61.08)
	Large (n=40)	5(12.50)	3(7.50)	5(12.50)	27(67.50)
	Pooled (n=180)	19(10.67)	16(8.8)	39(21.67)	106(58.50)

Figure in parentheses indicate percentage.

The variation of MEFL in SS was found significant (at 1% level of significance) between the extension systems and non significant between their categories (Table 8.24). Further, the variation of the same between the categories in each extension system was also found to be non significant (Table 8.25). It leads to inference that the farmers from the adopted villages of NDRI and SDAH, had better linkage with their EP with respect to supply and services, compared to the farmers from the adopted villages of HAU. The *t* value for the difference of significance between the perceived strength of functional linkage in SS by the EP and DFs was found to be significant and the gap in perception was worked out to be quite low (25.22%) as compared to the other parameters of functional linkage between them (Table 8.26). This led to the conclusion that both extension personnel and dairy farmers were better linked with respect to supply and services than the other parameters of functional linkage.

8.2.2.5 *Linkage in Training*

Training was identified yet another dimension of functional linkage between EP and DFs. In the present investigation, this was measured as the extent of joint involvement of both EP and DFs in all aspects of training, *viz.*, training need assessment, course curriculum development, conducting the training programmes and evaluation of the training programmes. The response was elicited separately and their categorization was done based on the MEFL in training.

A perusal of Table 8.15, reveals that majority (51.06%) of the sampled EP were in "weak" linkage category, followed by 23.40 per cent in "moderate" strength category. An equal percentage (12.77%) of them, however, were in the "no linkage" and "strong" linkage category. Extension system wise also, a similar trend was noted. Compared to NDRI and SDAH, more number of EP from HAU (66.67%) were under "weak" linkage category. The overall mean extent of linkage in training was computed to be 20.21, 38.43 and 28.54 per cent in case of NDRI, HAU and SDAH personnel, respectively (Table 8.16). This shows that HAU had taken a good lead to other extension systems as far as its functional interaction with dairy farmers on the parameter of training is concerned. The variation of MEFL in all the three system was statistically found to be non significant (Table 8.17).

Farmers' distribution on the basis of their perception of MEFL in training with EP was studied. As high as 79.44 per cent of them perceived "no linkage" with EP in training. The similar trend was noticed for the farmers of three systems as well as the categories (Table 8.22). The average extent of functional linkage on overall basis for farmers was worked out to be as low as 2.20, 1.59 and 0.93 per cent in case of NDRI, HAU and SDAH, respectively. Category wise these values were 0.75, 2.63 and 1.35 per cent for small, medium and large farmers, respectively (Table 8.23). The variation of extent of linkage between the category and between the system was found non significant (Table 8.24) and also the variation of the same among the categories within all the three systems was statistically non significant (Table 8.26). This led to inference that irrespective of the extension systems and farmers' categories, farmers and EP were poorly linked with each other with respect to training. Even the little claim by EP (29.37%) was badly negated by the dairy farmers (9.99%) as evident by the significant *t* value for the difference of mean of EFL in training between them. The percentage gap in linkage was also highest (65.98%) in case of this parameter (Table 8.26).

Table 8.22 : Frequency distribution of dairy farmers sampled from the selected extension systems based on their extent of linkage in Training with extension personnel

Extension Systems	*Category of Dairy Farmers*	*Extent of Linkage in Training*			
		No Linkage (0)	*Weak Linkage (<11.89)*	*Moderate Linkage (11.89 to 31.11)*	*Strong Linkage (>31.11)*
NDRI	Small (n=21)	15(71.43)	6(28.57)	–	–
	Medium (n=27)	20(74.07)	4(14.82)	3(11.11)	–
	Large (n=12)	8(66.67)	3(25.00)	1(8.33)	–
	Pooled (n=60)	43(71.67)	13(21.67)	4(6.66)	–
HAU	Small (n=22)	20(90.91)	2(9.09)	–	–
	Medium (n=25)	19(76.00)	6(24.00)	–	–
	Large (n=13)	11(84.62)	1(7.69)	1(7.69)	–
	Pooled (n=60)	50(88.33)	9(15.00)	1(1.67)	–
SDAH	Small (n=20)	18(90.00)	2(10.00)	–	–
	Medium (n=25)	17(68.00)	8(32.00)	–	–
	Large (n=15)	15(100.00)	–	–	–
	Pooled (n=60)	50(83.33)	10(16.67)	–	–
Overall	Small (n=63)	53(84.13)	10(15.84)	–	–
	Medium (n=77)	56(72.73)	18(23.37)	3(3.90)	–
	Large (n=40)	34(85.00)	4(10.00)	2(5.00)	–
	Pooled (n=180)	143(79.44)	32(17.78)	5(2.78)	–

Figure in parentheses indicate percentage.

8.2.2.6 *Overall Functional Linkage*

Based on the response elicited from both extension

personnel (EP) and dairy farmers (DFs) on the identified indicators of functional linkage between them, mean extent of overall functional linkage (MEOFL) value was obtained by taking average of the values of the selected indicators. Accordingly, categorization was done separately. The range of linkage strength for both the categories of respondents has been given as follows:

Linkage Strength	*Range for Extension Personnel*	*Range for Dairy Farmers*
No linkage	0	0
Weak linkage	Less than 28.75	Less than 11.88
Moderate linkage	28.75 to 60.45	11.88 to 31.11
Strong linkage	More than 60.45	More than 31.11

On the basis of these values, both EP and DFs were distributed separately under different linkage strength category. As evident from Table 8.5 that as high as 72.34 per cent of the pooled EP were coming under "moderate" strength of their functional linkage with the dairy farmers. This was followed by about 15 and 13 per cent of the EP falling under "weak" and "strong" linkage, respectively, with the dairy farmers. Extension system wise, a similar trend could be noted from the same table. It was heartening to note that none of the respondent from each selected extension system was in the "no linkage" category. It is also apparent from this table, that relatively more number of EP from HAU (26.67%) were in the "weak" linkage category than SDAH (18.75%). From NDRI, most of the EP were distributed in "moderate" (81.25%) and "strong" (18.75%) linkage category. The mean extent of overall functional linkage (MEOFL) for the pooled EP was computed to be 43.85 per cent. The values of MEOFL for the EP from NDRI, HAU and SDAH were 47.48, 38.81 and 45.25 per cent, respectively (Table 8.16). The variation of the values of MEOFL between the three selected extension systems was found statistically non significant (Table 8.17).

In case of sampled dairy farmers, most of them (48.89%) were found to be in "moderate", followed by 23.89 per cent in "weak", 17.22 per cent in "strong" and 10 per cent in "no" linkage categories. Except from HAU, dairy farmers from NDRI and SDAH displayed the similar trend (Table 8.18). From the same table, it is evident that about 30 per cent of the farmers from HAU were in "no" linkage category, followed by 35 per cent in "weak" linkage. Hence, more than 65 per cent of the farmers from HAU showed "no" to "weak" extent of interaction with their EP. The value of MEOFL for the farmers from NDRI, HAU and SDAH were computed to be 23.48, 9.77 and 24.78 per cent, respectively. Category wise, these values were 14.55, 21.27 and 21.91 per cent, respectively, for small, medium and large dairy farmers (Table 8.23).

Table 8.23 : Average extent of functional linkage (%) of dairy farmers with the extension personnel of selected extension system on various parameters.

Extension System	*Categories of Dairy Farmers*	*Parameters of Functional Linkages*					
		Communication	*Planning & Decisions making*	*Implementation & Evaluation*	*Supply and Services*	*Training*	*Overall*
NDRI	Small (n=21)	22.04	2.77	12.57	42.85	1.27	16.24
	Medium (n=27)	33.55	5.46	22.39	62.04	2.84	25.01
	Large (n=12)	41.52	9.03	26.28	67.36	2.50	29.20
	Pooled (n=60)	32.37	5.75	20.41	57.41	2.20	23.48
HAU	Small (n=22)	6.99	4.92	8.28	8.33	0.30	5.77
	Medium (n=25)	14.59	12.32	16.76	18.00	2.93	12.93
	Large (n=13)	12.96	10.89	19.81	9.62	1.54	10.61
	Pooled (n=60)	11.51	9.38	14.95	11.98	1.59	9.77
SDAH	Small (n=20)	25.33	2.58	23.99	59.99	0.67	21.65
	Medium (n=25)	29.78	10.67	25.20	65.67	2.13	25.87
	Large (n=15)	34.19	10.00	27.17	67.20	0.00	26.81
	Pooled (n=60)	29.77	7.75	25.39	64.29	0.93	24.78
Overall	Small (n=63)	18.12	3.42	14.93	37.06	0.75	14.55
	Medium (n=77)	25.97	7.02	21.45	48.57	2.63	21.27
	Large (n=40)	29.53	9.97	24.42	48.06	1.35	21.91
	Pooled (n=180)	24.54	6.80	20.27	44.56	1.58	1934

The values of MEOFL varied significantly (at 5% level of significance) between the selected three systems of extension. However, between the categories of the dairy farmers, this variation was not significant (Table 8.24). Further, the variation of MEOFL between the farmers of various categories under each selected extension system was found to be non significant (Table 8.25). It could be hence inferred that the farmers of HAU extension system was significantly poor linked with their EP compared to NDRI and SDAH systems. The effect of category of dairy farmers on their extent of interaction with the EP was non significant. The values of MEOFL of EP (44.60%) and DF (21.50%) was found to be significantly different and the perceived gap in linkage between them was computed to the extent of 51.79 per cent (Table 8.26).

Table 8.24 : Last-squares ANOVA for the means of functional linkage of farmers of various categories with extension personnel of the selected extension systems.

Sl. No.	*Parameters of functional linkage*	*Effects*	*SS*	*MSS*	*F-Value*
1.	Communication (2, 2 and 156)	Systems	7.2759	3.6379	46.1874**
		Categories	1.6774	0.8387	10.6483**
2.	Planning and decision making (2, 2 and 86)	Systems	1194.0453	597.0228	3.2778^{NS}
		Categories	675.9462	337.9731	1.8555^{NS}
3.	Implementation and evaluation (2, 2 and 155)	Systems	960.6853	480.3426	4.0592^{NS}
		Categories	2197.8296	1098.9148	9.1526^{NS}
4.	Supply and services (2, 2 and 154)	Systems	49768.2151	24884.107	110.5271*
		Categories	4121.2534	2060.6267	9.1526^{NS}
5.	Training (2, 2 and 25)	Systems	192.23917	96.11958	2.2197^{NS}
		Categories	23.3433	11.6716	0.2695^{NS}
6.	Overall (2, 2 and 156)	Systems	3160.4143	1580.2072	24.4393**
		Categories	1711.7867	855.89435	13.2374^{NS}

Figures in parentheses indicate degree of freedom.

* = Significant at 1 per cent level of significance, NS = Not significant.

** = Significant at 5 per cent level of significance.

The findings in above discussed subheads reveal that whereas SDAH mainly emphasised on supply and services, HAU system gave some attention to the participation of clients, though it was to the lesser extent, in the planning, deciding and implementing the extension activities. On the other hand, it could be observed that NDRI attempted to blend both educational and supply services in its extension approach. Very little to no participation of clients in extension activities was also reported by Gupta (1998) and Sharma and Rao (1998).

Table 8.25 : Least-squares ANOVA for the means of functional linkage of farmers of various categories within the selected extension systems

Sl. No.	*Parameters of functional linkage*	*Effects*	*SS*	*MSS*	*F-Value*
1.	Communication	NDRI (2 and 57)	3185.7610	1592.8805	20.3568**
		HAU (2 and 57)	234.2019	117.1009	0.9339**
		SDAH (2 and 57)	679.8002	339.9001	3.5718 NS
2.	Planning and decision making	NDRI (2 and 24)	104.5550	52.2775	1.4117NS
		HAU (2 and 38)	88.6333	44.3166	0.31050NS
		SDAH (2 and 27)	742.6804	371.3402	0.9935 NS
3.	Implementation and evaluation	NDRI (2 and 56)	1583.2004	791.6002	20.7555**
		HAU (2 and 31)	1293.7220	646.8610	3.3765**
		SDAH (2 and 57)	188.3947	94.1970	0.66625 NS
4.	Supply and services	NDRI (2 and 57)	6151.0005	3075.5002	13.4170 NS
		HAU (2 and 38)	113.9990	89.4995	0.3181 NS
		SDAH (2 and 57)	357.1088	178.5544	0.8229 NS
5.	Training	NDRI (2 and 8)	19.0279	9.5139	0.6963 NS
		HAU (2 and 6)	29.6370	14.8185	0.9379NS
		SDAH (2 and 8)	0.4004	0.4004	1.5465 NS
6.	Overall	NDRI (2 and 57)	1527.9885	763.9942	15.7560 NS
		HAU (2 and 38)	156.3861	78.1930	0.7740NS
		SDAH (2 and 57)	360.7765	180.3882	3.2634 NS

8.2.3 Extent of Functional Linkage Between Research Personnel and Dairy Farmers

The third dimension of research-extension farmers linkage was investigated in terms of extent of mutual and reciprocal linkage between research personnel and the dairy farmers. Only parameter-communication linkage was suggested by the judges to ascertain the extent of interaction between these two. Communication linkage was operationalized in terms of media/

channels jointly shared by RP and DFs. Their response was elicited separately, but on common instrument. Based on it, average extent of communication linkage (AECL) was computed and categorization was done.

Table 8.26 : Comparison of mean scores of extension personnel and farmers on selected parameters of functional linkages.

Sl. No.	Linkage Parameters (Indicators)	Extension Personnel (n=47)		Dairy Farmers (n=180)		$\overline{X}_1 - \overline{X}_2$	t-value	Gap(%) in Linkage
		Mean ($\overline{X}_1$)	S.D.	Mean ($\overline{X}_2$)	S.D.			
1	2	3	4	5	6	7	8	9
1.	Communication	51.65	15.11	26.82	12.31	25.13	11.47*	48.07
2.	Planning and decision making	41.59	19.22	16.50	13.94	25.09	8.44*	60.32
3.	Implementation and evaluation	44.31	15.71	22.37	11.62	21.94	10.25*	49.51
4.	Supply and services	66.70	25.70	49.88	23.53	16.82	4.18 NS	25.22
5.	Training	29.37	17.02	9.99	6.52	19.98	5.84**	65.98
6.	Overall	44.60	15.69	21.50	9.50	23.10	12.32*	51.79

* = Significant at 1 per cent level of significance. NS = Not significant

** = Significant at 5 per cent level of significance.

The distribution of research personnel according to their strength of communication linkage with the dairy farmers is presented in Table 8.27. From this table, it is evident that most of the research personnel (53.12%) had "moderate" strength, followed by "strong" linkage by 25.00 per cent of them. About 16 and 6 per cent of them were falling under "weak" and "no" linkage category, respectively. Research system wise, relatively more percentage (66.67%) of the personnel from HAU were under "moderate" linkage category than NDRI (41.18%). None of the research personnel from HAU expressed their absent linkage with the dairy farmers. However, this percentage was about 12 in case of the research personnel from NDRI.

Table 8.27 : Frequency distribution of research personnel sampled from the selected research systems based on their extent of communication linkage with farmers.

System	*Extent of Communication Linkage*				
	No. Linkage (0)	*Weak Linkage (8.02)*	*Moderate Linkage (8.02 to 38.16)*	*Strong Linkage (38.16)*	*Average Extent of Communication linkage*
NDRI (n=17)	2(11.76)	3(17.65)	7(41.18)	5(29.41)	22.06
HAU (n=15)	0(0.00)	2(13.33)	10(66.67)	3(20.00)	30.46
Pooled (n=32)	2(6.25)	5(15.63)	1753.12)	8(25.00)	26.26

Figures in parentheses indicate percentages.

Similarly, marginally greater percentage of personnel from NDRI (17.65%) were found to be in "weak" linkage category than the personnel from HAU (13.33%). Surprisingly, however, NDRI (29.41%) outperformed HAU (20.00%) when the percentage of personnel falling under "strong" linkage category was observed from the same table. When, the variation of extent of communication linkage scores were tested for significance between the two selected research systems, it was found to be non significant (Table 8.28). It is indicative of the fact that the variation in the distribution could be the matter of chance.

Table 8.28 : Least-squares ANOVA for the means of communication linkage of researchers with farmers of selected research systems.

Sl. No.	*Source*	*d.f.*	*S.S.*	*M.S.S.*	*F-Value*
1.	Communication linkage	1	43.3184	43.3184	0.1586^{NS}
2.	Error	30	69998.7461	233.2915	

NS = Non-significant

Farmers' view point about their extent of interaction with the research personnel was highly disheartening. None of the farmers from the sample reported to have any kind of interaction with the research personnel. Very few of them were found visiting the "Dairy Mela", but with little intention to interact with the scientist rather to enjoy mela and seek visual information. Surprisingly, also, farmers were unable to distinguish the researchers and their ideas of interaction with the farmers, even the later visited villages with field extension staff. The findings get the support from that of Gupta (1998).

8.3 Variables Affecting the Strength of Functional Linkage Between Research, Extension and Farmers

In order to identify the factors affecting the strength of functional linkage between research, extension and farmers, some of the variables were identified after review of literature. These variables were from broad areas of personal, psychological and organizational, for the sampled research and extension personnel. Similarly, for dairy farmers, personal, economic, psychological and communication variables were identified. These variables were put for subjective study and to identify the most important factors and their relative contribution objectively. Statistical tools as frequency, percentage, correlation, multiple regression and step wise multiple regression were employed for this purpose. Results have been presented and discussed under the following heads :

8.3.1 Variables Affecting the Strength of Functional Linkage Between Research and Extension Personnel

The selected variables studied have been subjectively described under the following subheads as the profile of research and extension personnel :

8.3.1.1 *Personal Variables*

8.3.1.1.1 Age

The findings in Table 8.29 indicates that most of the extension personnel (46.81%) were in the middle age group (between 41 to 51 years), followed by 36.17 per cent in young (upto 40 years) and 17.02 per cent in old (above 57 years) age category. The majority of sampled research personnel (50.00%), however, were in young age group, followed by equal percentage (25.00%) of them in middle and old age category.

Table 8.29 : Frequency distribution of research and extension personnel based on their personal variables

S. No.	*Variable*	*Categories*	*Scores*	*Research Personnel Pooled (n=32)*	*Extension Personnel Polled (n=47)*	*Overall (n=79)*	*Mean*	*S.D.*
1	2	3	4	5	6	7	8	9
1.	Age (yrs.)	Young	upto 40	16 (50.00)	17 (36.17)	33 (41.77)	42.73	8.03
		Middle	41 to 51	8 (25.00)	22 (46.81)	30 (37.97)		
		Old	Above 51	8 (25.00)	8 (17.02)	16 (20.26)		
2.	Education	Matriculation		0 (0.00)	2 (4.25)	2 (2.53)		
		Inter./ Diploma		0 (0.00)	5 (10.64)	5 (6.33)		
		Graduation		0 (0.00)	18 (38.30)	18 (22.78)		
		Masters		7 (21.87)	9 (19.15)	16 (20.25)		
		Ph.D.		25 (78.13)	13 (26.66)	38 (48.11)		
3.	Basic pay (Rs./ month)	Low	<2569	1 (3.13)	12 (25.53)	13 (16.46)	3960	1392
		Medium	2570-5352	24 (75.00)	26 (55.32)	50 (63.29)		
		High	>5352	7 (21.87)	9 (19.15)	16 (20.25)		
4.	Professional experience (years)	Low	<9	6 (18.75)	7 (14.89)	13 (16.46)	16.96	8.31
		Medium	9-17	12 (37.50)	16 (34.04)	28 (35.44)		
		High	>17	14 (43.75)	24 (51.07)	38 (48.10)		

(Contd...)

1 2	3	4	5	6	7	8	9
5. Cadre	Junior	1	8	12	20		
			(25.00)	(25.53)	(25.32)		
	Middle	2	15	24	39		
			(46.87)	(51.06)	(49.37)		
	Senior	3	9	11	20		
			(28.13)	(23.41)	(25.31)		
6. Training	No training		0	11	11		
received	1 month		(0.00)	(23.40)	(13.93)		
	training		0	16	18		
	3 months		(0.00)	(34.04)	(22.78)		
	training		0	14	14		
	5 months		(0.00)	(29.77)	(17.73)		
	training		2	2	4		
	8 months		(6.25)	(29.77)	(17.73)		
	training		5	1	6		
	More than		(15.63)	(2.12)	(7.59)		
	eight months		25	1	26		
	training		(78.12)	(2.12)	(32.91)		
7. Family size	Small	Upto 3	6	6	12	5.19	2.41
			(18.75)	(12.76)	(15.19)		
	Medium	4 to 6	21	30	51		
			(65.62)	(63.82)	(64.56)		
	Large	Above 6	5	11	16		
			(15.63)	(23.40)	(20.25)		
8. Family	Rural Agri.		9	10	19		
background			(28.12)	(21.27)	(24.05)		
	Rural-Non-		4	4	8		
	Agri.		(12.50)	(8.52)	(10.13)		
	Urban-Agri.		5	6	11		
			(15.63)	(12.76)	(13.92)		
	Urban-Non		14	27	41		
	Agri.		(43.50)	(57.45)	(51.90)		
9. Family type	Nuclear	1	25	26	51		
			(78.13)	(55.32)	(64.56)		
	Joint	2	7	21	28		
			(21.87)	(44.68)	(35.44)		

Figures in parentheses indicate percentage

8.3.1.1.2 Education

Regarding education, Table 8.29 indicates that majority of extension personnel (38.30%) were educated upto graduation, followed by 26.66, 19.15, 10.64 and 4.25 per cent were doctorate, masters, intermediate/diploma and matriculation passed, respectively. More number of extension workers from HAU were holding doctorate degree, whereas from SDAH and NDRI, majority of them were graduate. Research personnel, on the other hand, were mostly doctorate (78.13%), followed by masters (21.87%).

8.3.1.1.3 Basic Pay

Majority of the extension (55.32%) and research personnel (75.00%) were in medium pay scale category (Rs. 2570 5352 per month), followed by 25.53 and 19.15 per cent of extension personnel in low (less than Rs. 2569 per month) and high (more than Rs. 5352 per month) pay category, respectively. However, about 22 per cent of the research personnel were in high pay category, followed by only 3 per cent in low group.

8.3.1.1.4 Professional Experience

From the same table, it is evident that majority of the sampled extension and research personnel (48.10%) were having professional experience of more than 17 years, followed by 35.44 per cent of them having the same in between 9 to 17 years and 16.46 per cent in the low (less than 9 years) experience group. Almost similar trends could be observed for extension and research personnel separately in pooled case as well as organization wise.

8.3.1.1.5 Cadre

Findings of the table reveals that almost equal percentage (25%) of research and extension personnel were in junior and senior cadre in the organizational hierarchy. Rest of them (about 50%) were in middle cadre. It was also found that in HAU extension system, none of the EP were in junior cadre. It means that HAU has not employed junior cadre of EP and the middle level personnel only have to take care of field extension activities. Non availability of village level institutions with HAU extension system might be the reason for non employment of junior cadre of EP.

8.3.1.1.6 Training Received

From Table 8.29, it could be seen that majority of extension personnel (34.04%) had the exposure of one month in service training, whereas about 78 per cent of the research personnel were trained for the period above eight months. The findings in the table led to inference that research personnel were more exposed to in service training than the extension personnel.

8.3.1.1.7 Family Size

Majority of both research personnel (65.62%) and extension personnel (63.82%) had medium size of family, having 4 to 6 members. Table also reveals that whereas 23.40 per cent of the extension personnel had family members above 6; 18.75 per cent of the research personnel had family members upto 3.

8.3.1.1.8 Family Background

Majority of extension personnel (57.45%) and research personnel (43.50%) belonged to urban - non agriculture background. It was followed by 28.12 per cent of RP and 21.27 per cent of EP who had rural agriculture background.

8.3.1.1.9 Family Type

As high as 78.15 per cent of the RP and 53.32 per cent of EP belonged to nuclear families. The remaining percentage of them were from joint families.

8.3.1.2 *Psychological Variables*

8.3.1.2.1 Attitude of the Personnel

Attitude of the research and extension personnel was studied on six dimensions. The findings as presented in Table 8.30 reveals that majority of extension personnel and research personnel expressed "neutral attitude" with respect to work, working condition, co-worker, supervisor, organization and management.

With respect to "work", "supervisor" and "overall attitude", considerable percentage of research personnel from HAU, *i.e.*, 40.00, 66.67 and 53.33 per cent, respectively, had negative attitude. In case of extension personnel from NDRI, 31.25, 37.50 and 37.50 per cent of them had positive attitude with respect to "work", "organization" and "overall attitude". Similarly, a good percentage of extension personnel from SDAH, *i.e.*, 26.58, 40.43, 27.66 and 29.79 per cent of them had positive attitude with respect to "work", "organization", "management" and "overall attitude". From the above result, it could be inferred that extension personnel had relatively more positive feelings on the all dimensions of attitude than the research personnel.

8.3.1.2.2 Achievement Motivation

From the same table, it could be observed that as high as 76.59 and 65.63 per cent of the extension personnel and research personnel had average level of achievement motivation. Finding also shows that more number of research personnel (21.87%) had lower level of achievement motivation than the extension personnel (6.38%). This scenario was much pronounced in case of the personnel from HAU.

8.3.1.2.3 Value Orientation

Findings in the same table reveal that 65.63 and 57.45 per cent of research and extension personnel had medium level of value orientation. An equal percentage of both of them (about 25%) were found to have high value orientation. It means that with the change of time, these personnel are moving towards modern value system.

8.3.1.2.4 Job Satisfaction

Majority of the research personnel (71.87%) and extension personnel (61.70%) had medium level of job satisfaction. This was followed by 21.28 and 17.02 per cent of the EP having low and high level of job satisfaction, respectively. In case of research personnel, however, relatively more percentage (18.76%) of them had higher level of job satisfaction than those (9.37%) with lower level of the same.

8.3.1.2.5 Morale

A cursory look of the same table, shows that majority of the extension personnel (62.43%) and research personnel (59.38%) had medium level of morale with respect to fairness of the policy of the organization. However, more number (26.04%) of RP were found in high morale category than the EP (17.02%).

Similarly, on the other four dimensions of morale studied, majority of both research and extension personnel were in average category. However, 34.38, 28.13, 28.12 and 25.00 per cent of research personnel were in low morale category with respect to "adequacy of leadership", "sense of participation in management", "sense of regard and identification" and "overall morale", respectively. Relatively more number of EP from HAU were having low morale level than the personnel from HAU. Extension personnel on the other hand, showed diverse response. Whereas, 38.29 per cent of them had high morale with respect to "sense of participation in management", almost equal percentage of them were in high and low category with respect to "fairness of policy" (about 17.02%), "adequacy of leadership" (14.89%) and "sense of regard and identification" (about 13%). Further, it was found that the EP from SDAH had relatively higher morale than those from NDRI and HAU.

8.3.1.3 *Organizational Variables*

8.3.1.3.1 Perception of Management (PM)

As indicated in Table 8.31, most of the research personnel (62.50%) and extension personnel (74.47%) were having medium (average) perception of the management of their organization. Relatively more number of research personnel (31.25%) were found to have low (poor) perception compared

Table 8.30 : Frequency distribution of research and extension personnel based on their psychological variables.

S. No.	Variable	Categories	Scores	Research Personnel (n=32)	Extension Personnel (n=47)	Overall (n=79)	Mean	S.D.
1	2	3	4	5	6	7	8	9
1.	**Attitude Towards**							
(a)	Work	Negative	< 6.48	8 (25.00)	6 (12.76)	14 (17.72)	7.70	2.41
		Neutral	6.48 - 8.93	18 (56.25)	26 (55.32)	44 (55.70)		
		Positive	> 8.93	6 (18.75)	15 (31.92)	21 (26.58)		
(b)	Working condition	Negative	<3.73	5 (15.63)	4 (8.51)	9 (11.39)	4.54	0.81
		Neutral	3.73–5.35	22 (68.75)	37 (78.72)	59 (74.68)		
		Positive	>5.35	5 (15.62)	6 (12.77)	11 (13.93)		
(c)	Co-workers	Negative	<3.84	6 (18.75)	6 (12.77)	12 (15.19)	4.92	1.08
		Neutral	3.84-6.00	22 (68.75)	38 (80.85)	60 (75.94)		
		Positive	>6.00	4 (12.50)	3 (6.38)	7 (8.87)		
(d)	Supervisor	Negative	<7.12	12 (37.50)	4 (8.52)	16 (20.25)	9.62	2.50
		Neutral	7.12-12.12	16 (50.00)	37 (78.72)	53 (67.09)		
		Positive	>12.11	4 (12.50)	6 (12.76)	10 (12.66)		
(e)	Organization	Negative	<1.73	2 (6.25)	3 (6.38)	5 (6.33)	2.27	0.57
		Neutral	1.73-2.87	21 (65.63)	25 (53.19)	46 (58.23)		
		Positive	>2.87	9 (28.12)	19 (40.43)	28 (35.44)		

(Contd...)

1 2	*3*	*4*	*5*	*6*	*7*	*8*	*9*
(*f*) Management	Negative	<6.20	6 (18.75)	10 (21.28)	16 (20.25)	7.29	1.09
	Neutral	6.20-8.38	24 (75.00)	24 (51.06)	48 (60.76)		
	Positive	>8.38	2 (6.25)	13 (27.66)	15 (18.99)		
(*g*) Overall	Negative	<31.77	13 (40.63)	4 (8.51)	17 (21.52)	36.97	5.20
	Neutral	31.77-41.26	14 (43.75)	29 (61.70)	43 (54.43)		
	Positive	>41.26	5 (15.62)	14 (29.79)	19 (24.05)		
2. Achievement motivation	Low	<21.34	7 (21.87)	3 (6.38)	10 (12.66)	23.90	2.56
	Medium	21.35-26.46	21 (65.63)	36 (76.59)	57 (72.15)		
	High	>26.46	4 (12.50)	8 (17.03)	12 (15.19)		
3. Value orientation	Low	<27.62	3 (9.37)	8 (17.03)	11 (13.92)	30.25	2.62
	Medium	27.62-32.87	21 (65.63)	27 (57.45)	48 (60.76)		
	High	>32.87	8 (25.00)	12 (25.53)	20 (25.32)		
4. Job satisfaction	Low	<58.59	3 (9.37)	10 (21.28)	13 (16.46)	66.92	8.33
	Medium	58.59-75.25	23 (71.87)	29 (61.70)	52 (65.82)		
	High	>75.25	6 (18.76)	8 (17.02)	14 (17.72)		
5. **Morale:**							
(*a*) Fairness of policies	High	<6.34	6 (26.04)	8 (17.02)	14 (17.72)	9.70	3.36
	Medium	6.34-13.06	19 (59.38)	30 (63.83)	49 (62.03)		

(Contd...)

1 2	3	4	5	6	7	8	9
	Low	>13.06	7 (21.88)	9 (19.15)	16 (20.25)		
(*b*) Adequacy of leader-ship	High	<8.83	4 (12.50)	7 (14.89)	11 (13.92)	11.70	2.87
	Medium	8.83-14.57	17 (53.12)	33 (70.22)	50 (63.29)		
	Low	>14.57	11 (34.38)	7 (14.89)	18 (22.79)		
(*c*) Sense of participa-tion in management	High	<7.07	6 (18.75)	18 (38.29)	24 (30.38)	10.39	3.32
	Medium	7.09-13.71	17 (53.12)	19 (40.42)	36 (45.57)		
	Low	>13.71	9 (28.13)	7 (14.89)	16 (20.25)		
(*d*) Sense of regard and identification	High	<6.45	2 (6.25)	7 (14.89)	9 (11.39)	9.02	2.56
	Medium	6.45-11.58	21 (65.63)	34 (72.34)	55 (69.63)		
	Low	>11.58	9 (28.12)	6 (12.77)	15 (18.98)		
(*e*) Overall moral	High	<32.14	5 (15.63)	9 (19.15)	14 (17.72)	12.70	2.42
	Medium	32.14-54.16	19 (59.37)	34 (72.34)	53 (67.09)		
	Low	>54.16	8 (25.00)	4 (8.51)	12 (15.19)		

Figures in parentheses indicate percentage.

to the extension personnel (2.12%). Situation was poorer in case of HAU where about 47 per cent of the sampled research personnel perceived low (poor) management in their organization.

The response of extension and research personnel was also taken on some of the items of the PM scale and based on their

Table 8.31 : Frequency distribution of research and extension personnel based on their perception of management of the organization/ department

S.No.	Variables	Categories	Scores	Research Personnel Polled (n=32)	Extension Personnel Polled (n=47)	Overall (n=79)
1	2	3	4	5	6	7
	Overall Perception of Management (PM)	Low (Poor)	<48.53	10 (31.25)	1 (2.12)	11 (13.92)
		Medium (Average)	48.53-67.15	20 (62.50)	35 (74.47)	55 (69.62)
		High (Good)	>67.15	2 (6.25)	11 (23.41)	13 (16.46)
(a)	Farmers' need based planning and formulation of research and extension projects/programmes	Strongly agree	5	2 (6.25)	12 (25.53)	14 (17.72)
		Agree	4	12 (37.50)	23 (48.94)	35 (44.30)
		Undecided	3	5 (15.63)	3 (6.38)	8 (10.13)
		Disagree	2	11 (34.37)	8 (17.02)	19 (24.05)
		Strongly disagree	1	2 (6.25)	1 (2.13)	3 (2.53)
(b)	Decision with respect to research and extension activities are executed without much delay	Strongly agree	5	0 (0.00)	9 (19.15)	9 (3.80)
		Agree	4	10 (31.25)	23 (48.94)	33 (41.77)
		Undecided	3	6 (18.75)	5 (10.64)	11 (13.92)
		Disagree	2	16 (50.00)	10 (21.27)	26 (32.91)
		Strongly disagree	1	0 (0.00)	0 (0.00)	0 (0.00)
(c)	Authorities and power are centralised in this organization	Strongly agree	1	6 (18.75)	5 (10.64)	11 (13.92)
		Agree	2	14 (43.75)	17 (36.17)	31 (39.24)

(Contd...)

1 2	3	4	5	6	7®
	Undecided	3	5 (33.33)	10 (21.28)	15 (18.99)
	Disagree	4	7 (21.87)	14 (29.79)	21 (26.58)
	Strongly disagree	5	0 (0.00)	1 (2.12)	1 (1.27)
(*d*) Jealousy and leg pulling is prevalent in this organization	Strongly agree	1	4 (12.50)	1 (2.12)	5 (6.34)
	Agree	2	11 (34.38)	17 (36.17)	28 (35.44)
	Undecided	3	6 (18.75)	12 (25.53)	18 (22.78)
	Disagree	4	9 (28.12)	12 (25.53)	21 (26.58)
	Strongly disagree	5	2 (6.25)	5 (10.65)	7 (8.86)
(*e*) Farmers' programme is not effective because of poor coordination	Strongly agree	1	1 (3.12)	5 (10.64)	6 (7.59)
	Agree	2	11 (34.38)	5 (10.64)	16 (20.26)
	Undecided	3	4 (12.50)	8 (17.02)	12 (15.19)
	Disagree	4	11 (34.38)	19 (40.42)	30 (37.97)
	Strongly disagree	5	5 (15.62)	10 (21.28)	15 (18.99)
(*f*) The communication network between farmers and institute is adequate	Strongly agree	5	3 (9.37)	12 (25.53)	15 (18.98)
	Agree	4	12 (37.50)	19 (40.42)	31 (39.24)
	Undecided	3	7 (21.87)	6 (12.76)	13 (16.46)
	Disagree	2	10 (31.25)	8 (17.02)	18 (22.78)
	Strongly disagree	1	0 (0.00)	2 (4.26)	2 (2.54)

Figures in parentheses indicate percentage.

response, frequency distribution was made and presented in the same table. A cursory look of the table reveals that a considerable percentage of extension personnel (44.30) and research personnel (37.50) agreed with the point that planning and formulation of research and extension projects were based on farmers' need. However, a diverse response was noted with respect to the timely execution of research and extension activities. Whereas, half of the RP disagreed with this fact, about 49 per cent of the EP expressed their agreement regarding timely execution of the field extension activities. Further, with respect to the perception items, *viz.*, "centralised authority and power" and "prevalence of jealousy and leg pulling in the organization", it was found that majority of EP and RP showed their agreement. The response was more frequent in case of HAU compared to NDRI. Similarly, majority of RP (34.38%) perceived that farmers' programme was not effective because of poor coordination in the organization. Contrastingly, however, most of the EP (40.92%) negated this point. With respect to adequacy of network between farmers and institute, about 38 and 40 per cent of the RP and EP expressed their agreement. Substantially, lesser per cent of RP from NDRI (29.41%) agreed with this point compared to those from HAU (46.67%).

From the preceding findings, it could be observed that planning and formulation of research and extension projects/ programmes were claimed to based on farmers' need, their execution was delayed. This may probably be due to prevalence of jealously and leg pulling in the organization. Moreover, centralized power and authority might have further contributed in delayed execution of those activities. It was also realised that though a good communication network between farmers and institute existed, but poor coordination was felt by the research personnel in making the farmers' programme effective. This clearly leads to conclusion that existing structural linkage mechanisms were either non-functional or they were inadequate.

8.3.1.3.2 Organizational Climate

Organizational climate was identified as yet another important organizational variable. This was studied with respect to ten dimensions of organizational climate. The response of both RP and EP was elicited on bi-polar statements. The findings are contained in Table 8.32 and discussed below :

Table 8.32 : Frequency distribution of research and extension personnel based on their perception about organizational climate.

S.No.	Variables	Categories	Scores	Research Personnel Polled (n=32)	Extension Personnel Polled (n=47)	Overall (n=79)
1	2	3	4	5	6	7
1.	Conformity	Low	<4	2 (6.25)	8 (17.02)	10 (12.66)
		Medium	4-8	20 (62.50)	25 (53.19)	45 (56.96)
		High	>8	10 (31.25)	14 (29.79)	24 (30.38)
2.	Responsibility	Low	<4	7 (21.87)	8 (17.02)	15 (18.99)
		Medium	4-8	24 (75.00)	27 (57.45)	51 (64.56)
		High	>8	1 (3.13)	12 (25.53)	13 (16.45)
3.	Standards	Low	<4	7 (21.88)	11 (23.40)	18 (22.78)
		Medium	4-8	22 (68.75)	29 (61.70)	51 (64.56)
		High	>8	2 (9.37)	5 (14.90)	7 (12.66)
4.	Reward	Low	<4	10 (31.25)	14 (29.79)	24 (30.38)
		Medium	4-8	20 (62.50)	28 (59.57)	48 (60.76)
		High	>8	2 (6.25)	5 (10.64)	7 (8.86)
5.	Organizational clarity	Poor	<4	2 (6.25)	8 (17.02)	10 (12.66)
		Medium	4-8	25 (78.12)	29 (61.70)	54 (68.35)
		High	>8	5 (15.63)	10 (21.28)	15 (18.99)

(Contd...)

1 2	3	4	5	6	7
6. Warmth and support	Low	<4	5 (15.63)	14 (29.79)	19 (24.05)
	Medium	4-8	26 (81.25)	26 (55.32)	52 (65.82)
	High	>8	1 (3.12)	7 (14.89)	8 (10.13)
7. Leadership	Low	<4	12 (37.50)	14 (29.79)	26 (32.92)
	Medium	4-8	18 (56.25)	29 (61.70)	47 (59.49)
	High	>8	2 (6.25)	4 (8.51)	6 (7.59)
8. Overall organizational climate	Low	>32.14	6 (18.75)	9 (19.15)	15 (18.98)
	Medium	32.14-54.16	25 (78.13)	26 (55.32)	51 (64.56)
	High	>54.16	1 (3.12)	12 (25.53)	13 (16.56)

A critical look of the table indicates that more than 50 per cent of both RP and EP perceived medium level of overall organizational climate. Extension personnel (25.53%) perceived higher level of overall organizational climate more frequently than the research personnel (3.12%). Further, it is also evident that on all the dimensions of organizational climate studied, more than half of the sampled personnel were in medium level of perception. With respect to the dimension conformity, about 30 per cent of both EP and RP had perceived it high. Similarly, about 26 per cent of EP felt high sense of responsibility of the job as compared to only 3.13 per cent of RP who hold the similar perception. "Standard of job" was also perceived low by considerable percentage of RP (21.88%) and EP (23.40%). Their perception regarding "reward system", and "warmth and support" in the organization was in the similar line. With respect to the most important aspect of organizational climate, *i.e.*, leadership, about 38 and 30 per cent of RP and EP, respectively, felt it as low.

Table 8.33 : Frequency distribution of research and extension personnel based on their response on the factors external to the organization and affecting the research and extension activities.

Sl. No.	*External Factors*	*Response*	*Research Personnel (n=32)*	*Extension Personnel (n=47)*	*Overall (n=79)*
1	2	3	4	5	6
1.	Government policy	Yes	7 (21.88)	10 (21.28)	17 (21.53)
		No	16 (50.00)	31 (65.96)	47 (59.49)
		Uncertain	7 (21.78)	5 (10.63)	12 (15.19)
		Not relevant	2 (6.24)	1 (2.13)	3 (3.79)
2.	Farmers' organization	Yes	0 (0.00)	5 (10.64)	5 (6.33)
		No	7 (21.87)	27 (57.45)	34 (43.04)
		Uncertain	14 (43.75)	12 (25.53)	26 (32.91)
		Not relevant	11 (34.38)	3 (6.38)	14 (17.72)
3.	Foreign agencies	Yes	3 (9.38)	0 (0.00)	3 (3.79)
		No	18 (56.25)	9 (19.15)	27 (34.18)
		Uncertain	10 (31.25)	20 (42.55)	30 (37.97)
		Not relevant	1 (3.12)	18 (38.30)	19 (24.06)
4.	Private organizations	Yes	0 (0.00)	0 (0.00)	0 (0.00)
		No	9 (28.13)	20 (42.55)	29 (36.71)
		Uncertain	8 (25.00)	15 (31.92)	23 (29.11)
		Not relevant	15 (46.87)	12 (25.53)	27 (34.18)

From the above findings, it could be inferred that both the selected research and extension organizations/departments had more of bureaucratic bent. This fact is also evident from high degree of conformity to the rules and regulation, inadequacy of immediate leadership and improper reward system. These might be the reasons for poor perception of the personnel towards the management (Table 8.31). This also led to inference that poor extent of functional linkage between research and extension was the dysfunctional consequence of poor organizational climate and poor perception of organization management.

8.3.1.3.3 External Factors

Based on the literature, some of the factors external to the organization were identified which could constrain research and extension activities. These factors were: government policy, farmers' organization, foreign agencies and private organizations. Response of EP and RP was ascertained on four points of "Yes", "No", "Uncertain" and "Not relevant". Based on the frequency of response, they are distributed in Table 8.33.

Regarding "government policy", about 66 per cent of EP and half of the RP found it not affecting their works. However, about 21 per cent of both of them found the policy of government affected research and extension activities. About 15 per cent of them were uncertain on this aspect.

Similarly, about 57 per cent of EP found that there was no effect of farmers' organization on the extension activities. This was followed by 25.53 and 10.64 per cent of EP who were uncertain about its effect on extension work, respectively. Majority of research personnel, on the other hand, were uncertain (43.75%) about it and also found it "not relevant" (34.38%). The effect of foreign agencies on research was perceived "no" to "uncertain" by majority of the RP (87.50%). Extension personnel also found it "uncertain" (42.55%) and "not relevant" (38.30%). Similarly, whereas about 47 per cent of RP considered private organizations "not relevant" to their work, majority of EP (43%) found them as the factors not affecting their extension work.

From the above findings, it could be argued that the selected research and extension systems had almost nil influence of external factors. The reason behind the same might be the belonging of selected systems to the public sector. It could also be inferred that both research and extension activities in the areas studied were little directed by farmers' organization and foreign agencies. Similarly, the affect of private organizations was also not found. Hence, it could be aptly concluded that the research and extension activities were being done by the selected organization/departments with little to nil affect from the external environment.

8.3.1.3.4 Goals of the Organization

Some of the general items were identified which acted as the common goal to both research and extension organizations in relation to dairy development. These goals were scored by the sampled personnel and separate ranking was done for RP and EP. From the findings as contained in Table 8.34, it could be observed that EP gave first rank to the commitment of department to improve the indigenous breeds of cattle and buffalo in the area under their operation. This item was found to have been accorded second rank by the research personnel. Research personnel, however, gave first rank to their concern to improve milk yield of cows and buffalo. This was given second rank by the extension personnel. The goals like "strong commitment to bring down the mortality rate in cattle and buffalo by providing better veterinary services", "cattle owners should get required inputs in time and at appropriate price" and "to improve the socio economic status of weaker section through dairying" were given third, fourth and fifth rank, respectively, by the extension personnel. These goal items, however, were positioned on sixth, tenth and eighth rank, respectively, as per the response of RP. According to research personnel, their third important goals was "to conduct more of basic and applied researches". This was followed by "to educate the farmers about breeding, feeding, health care and management aspects" and "fodder development" as the third, fourth and fifth important goals, respectively. These items were

Table 8.34 : Rank order of the goals of the organization/department as perceived by the research and extension personnel

Sl. No.	Goals	*Scores (Rank) of Extension Personnel (N=47)*	*Scores (Rank) of Research Personnel (N=32)*	*r-Value*
1.	Department is strongly committed to improve the indigenous breeds of cattle and buffalo in the area under its operation	15.24(I) (I)	14.13 (II)	
2.	Department is concerned to improve the milk yield of cows and buffalo	14.20 (II)	16.00 (I)	
3.	Department is strongly committed to bring down the mortality rate in cattle and buffalo by providing better veterinary facility	13.00 (III)	10.01 (VI)	
4.	Department also believe that the cattle owners should get required inputs in time and at appropriate price	12.31 (IV)	2.01 (X)	
5.	Organization aims at improving the socio-economic status of weaker section through dairying	10.05 (V)	8.89 (VII)	
6.	Providing technical inputs and extension services to farmers is the main aim of department	9.54 (VI)	5.32 (IX)	
7.	Department believes that the farmers should be educated about the aspects, *viz.*, breeding, feeding, health care and management	8.50 (VIII)	12.62 (IV)	
8.	Fodder development is one of the objectives of department	8.00 (IX)	11.22 (V)	
9.	Department is committed to generate employment for the educated unemployed through dairying	9.16 (VII)	6.64 (VIII)	
10.	Department is concerned with more of basic and applied researches	0.00(X)	13.26(III)	0.401^{NS}

NS = Non significant.

accorded tenth, eighth and ninth rank, respectively, by the extension personnel.

From the above findings, it is obvious that there was a sharp line of difference in perception of the goals of organization by RP and EP. This fact is further validated with very low and non significant value of rank correlation coefficient. This differential perception in the goals of the organization might be due to the fact that both research and extension are operating in isolation. Very poor extent of functional linkage between them might be the result of such deviation in the perception. This also led to inference that mere installation of structural linkage mechanism would do little in improving the functional linkage unless the goals of the organization/department are similarly perceived by both RP and EP.

8.3.1.4 Variables Affecting Functional Linkage Between Research and Extension

In order to identify the variables affecting the extent of functional linkage between research and extension objectively, zero order correlation, multiple regression and step wise regression were employed. The findings so obtained are discussed below :

From Table 8.35, it could be observed that variables like, cadre, age, educational qualification, professional experience were significantly associated (at 1%) with the extent of communication linkage (ECL) between research and extension. The variable job satisfaction had significant co-variance with the ECL at 5 per cent level of significance. Rest of the selected variables were found having non significant correlation. Among all the significantly associated variables, age had negative and significant association. When the set of independent variables were subjected for multiple regression analysis, only two variables, *viz.*, educational qualification and job satisfaction were found to have significant effect on the dependent variable - ECL at 5 per cent level of significance. The variation in the ECL was predicted to the extent of 50 per cent when all fourteen variables were taken together. A significant *F* value for corresponding R^2 value indicated a good fit of the equation

(Table 8.36). On step wise regression, however, only three variables retained in the final model and these jointly predicted 48 per cent variation in the ECL (Table 8.37). Among the three, job satisfaction and educational qualif ation were contributing significantly. The share of job satisfaction, educational qualification and cadre was about 20, 18 and 10 per cent, respectively.

From the findings, it could be concluded that only highly qualified personnel who were also satisfied with their job had good extent of communication linkage. This also mean that the extension personnel from lower cadre and with lower educational qualification did not have good linkage with the research personnel. From the same table, it could be seen that only cadre was significantly (at 5% level of significance) associated with the extent of linkage in collaborative professional activities (ELCPA) between research and extension. On multiple regression analysis also, only cadre was found to have significant influence on the dependent variable ELCPA. All the selected fourteen variables could predict only to the extent of 35 per cent in the variation of dependent variable (Table 8.36). The non significant F value for R^2 value indicated that the model did not prove the good fit. The findings led to conclusion that parameter like collaborative functional activities was least affected by the profile of selected personnel, their psychological variables and organizational variables as perceived by them.

Similarly, with the third parameter, two variables namely, cadre and educational qualification were found significantly associated (Table 8.3.1.7). However, none of the selected variables were having causal influence on it (Table 8.35). The R^2 value was only to the extent of 32 per cent.

Regarding implementation and evaluation, only cadre of the personnel was found significantly associated with the EFL on this aspect (Table 8.35). From multiple regression analysis, two psychological variables namely, attitude and value orientation of the personnel had significant causative effect on the dependent variable. Still, the R^2 value was as low as 25 per cent.

Table 8.35 : Zero order correlation coefficient of selected variables with the extent of functional linkage between research and extension personnel.

S. No.	Variables	Parameters of Functional Linkage						
		Communi-cation	Collabora-tive Pro-fessional Activities	Planning & Decision Making	Implemen-tation & Evaluation	Supply and Services	Training	Overall
1.	Cadre (X_2)	0.4192*	0.2643**	0.3705*	0.2387**	0.3366*	0.0115	0.2237**
2.	Age (X_3)	-0.4013*	-0.1513	-0.1820	-0.2063	-0.2574**	0.0678	-0.1972
3.	Educational qualification (X_4)	0.5431*	0.2277	0.4624*	0.1757	0.3342*	-0.0818	0.2121
4.	Professional experience (X_5)	0.3225*	0.1036	0.0805	0.1441	0.1674	0.632	0.1633
5.	Training received (X_6)	0.0952	0.0721	0.1213	0.0211	0.0392	-0.0260	0.4133*
6.	Attitude (X_7)	0.0232	0.0188	0.0137	0.0988	0.1135	0.2467**	0.3628*
7.	Family background (X_8)	-0.1417	-0.0437	-0.1402	-0.0458	-0.0598	0.0031	-0.0809
8.	Achievement motivation (X_9)	0.1001	0.0240	0.0112	0.0138	0.0264	0.0822	0.1103
9.	Value orientation (X_{10})	0.0050	-0.0649	-0.0227	-0.1756	-0.0798	-0.0696	-0.0203
10.	Job satisfaction (X_{11})	0.2250**	0.0612	0.0177	0.0116	0.0435	0.2119	0.0740
11.	Morale (X_{12})	0.0017	0.0640	0.0256	0.0733	0.1768	-0.1168	-0.1795
12.	Perception of management (X_{13})	0.0814	0.0937	0.0884	0.0499	0.1253	0.0949	0.1828
13.	External environment (X_{14})	0.0341	-0.0646	0.0330	-0.0449	0.0019	0.0349	0.0805
14.	Organizational climate (X_{15})	0.0144	0.0662	0.0203	0.0338	0.0700	0.0884	0.2684**

* = Significant at 1 percent level of significance; ** = Significant at 5 percent level of significance.

Table 8.36 : Multiple regression coefficients of the selected independent variables of dairy farmers with their extent of functional linkage with extension personnel.

S. No.	Variables	Parameters of Functional Linkage													
		Communi-cation		Collabora-tive Pro-fessional Activities		Planning & Decision Making		Implemen-tation & Evaluation		Supply and Services		Training		Overall	
		b	t	b	t	b	t	b	t	b	t	b	t	b	t
1.	Cadre (X_2)	4.6134	0.8955	-8.2196	1.7346**	-5.5440	1.2990	-4.5113	1.3494	-2.6291	0.6494	-1.4709	0.3971	-8.7739	2.0335**
2.	Age (X_3)	-0.2058	0.3070	0.2502	0.4057	0.1352	0.2435	0.3570	0.8206	0.5088	0.9661	0.1472	0.3055	0.5374	0.9573
3.	Educational qualification (X_4)	0.8131	2.4080**	0.3619	0.1390	3.5800	1.5273	0.7085	0.3859	0.9446	0.4249	0.2854	0.1403	6.4137	2.7066*
4.	Professional experience (X_5)	0.3161	0.4929	-0.4211	0.7140	0.5630	1.0597	0.4045	0.9720	0.7941	1.5758	0.0969	0.2101	-0.8234	1.5331
5.	Training received (X_6)	1.1179	0.9016	0.9860	0.8645	0.8811	0.8577	0.3438	0.4272	0.6957	0.7140	-0.9977	1.1190	2.0942	2.0166**
6.	Attitude (X_7)	0.6574	1.3100	0.5018	1.0871	0.6348	1.5270	0.6364	1.9543**	0.0640	0.1623	0.6948	1.9255**	0.7291	1.7348**
7.	Family back-ground (X_8)	-0.1001	0.1411	-0.2329	0.3570	-0.4320	0.7351	0.1637	0.3556	-0.0321	0.0575	-0.1159	0.2272	-0.9115	1.5343
8.	Achievement motivation (X_9)	0.5220	0.6379	-0.0145	0.0193	-0.0063	0.0093	0.2486	0.4682	0.3792	0.5896	0.0219	0.0371	0.3140	0.4582
9.	Value orienta-tion (X_{10})	-0.4965	0.6432	-0.7263	1.0230	-0.6514	1.0186	-1.0166	2.0296**	-0.5942	0.9797	-0.4141	0.7461	-.2440	0.3774
10.	Job satisfaction (X_{11})	0.6550	2.5234**	0.1683	0.7047	0.0676	0.3142	0.0379	0.2250	0.1005	0.4929	0.1707	0.9146	0.0810	0.3728
11.	Morale (X_{12})	0.1494	0.4582	0.1022	0.3405	0.1583	0.5857	0.1655	0.7818	0.4431	1.7288**	-0.1183	0.5042	0.0694	0.2541

(contd...)

S. No.	Variables	Parameters of Functional Linkage													
		Communi-cation		Collabora-tive Pro-fessional Activities		Planning & Decision Making		Implemen-tation & Evaluation		Supply and Services		Training		Overall	
		b	*t*	*b*	*t*	*b*	*t*	*b*	*t*	*b*	*t*	*b*	*t*	*b*	*t*
12.	Perception of management (X_{13})	0.0559	0.1665	0.1779	0.5758	0.0264	0.0948	0.0886	0.4068	0.2101	0.7964	-0.2212	0.9166	0.2555	0.9087
13.	External environment (X_{14})	-0.0869	0.0862	-1.0983	1.1852	0.3684	0.4414	-0.9536	1.4585	0.3482	0.4398	0.2512	0.3468	-0.5976	0.7082
14.	Organizational climate (X_{15})	0.0763	0.3893	0.0730	0.4050	0.0437	0.2693	0.0255	0.2008	0.793	0.5150	0.0035	0.0251	0.1716	1.0452
	R^2 Value (*F* Value at 15 and 63 d.f.)	0.50 (4.22*)		0.35 (0.74^{NS})		0.32 (1.90^{NS})		0.25 (1.00^{NS})		0.24 (1.33^{NS})		0.20 (0.63^{NS})		0.40 (2.72**)	

* = Significant at 1 per cent level of significance; ** = Significant at 5 per cent level of significance.

Similarly, with respect to supply and services, the variables like, cadre and educational qualification were found to have significant association (at 1% level) with the extent of functional linkage and age was significantly co-varied at 5 per cent level of significance (Table 8.35). However, on regression analysis, none of these variables had significant influence, except the moral of the employee which had causative effect on the dependent variable at 5 per cent level of significance (Table 8.36). The R^2 value was low to the extent of 24 per cent.

The last parameter, training also showed the similar nature of association and cause and effect relationship with the selected antecedent variables. Only attitude was found to have significant association and affect on the extent of functional linkage in training (Table 8.35). The overall extent of functional linkage (OEFL) between research and extension was observed to be associated with several background variables. Whereas cadre and organizational climate were significantly associated with OEFL at 5 per cent level of significance, training received and attitude were significantly associated at 1 per cent level of significance.

Table 8.37 : Results of stepwise regression analysis (backward elimination method) of independent variables with the extent of communication linkage between research and extension personnel.

Sl. No.	*Independent Variables*	*b-Values*	*t-Values*	*Relative Contribution (%)*
1.	Cadre (X_2)	6.0116	1.3707^{NS}	9.78
2.	Educational qualification (X_4)	5.9503	3.0900*	18.22
3.	Job satisfaction (X_{11})	0.7200	3.4467*	20.00

* = Significant at 1 per cent level of significance;

NS = Non significant.

R^2 Value = 0.48; *F* Value = 16.59* with 3 and 75 d.f.

On multiple regression analysis, cadre, educational qualification and attitude of the personnel had significant influence of the dependent variable - OEFL. The R^2 value was 0.40 and it was significant at 5 per cent level of significance. It means that all the set of independent variables could predict only to the extent of 40 per cent in the variation of dependent variable. On stepwise regression analysis (Table 8.38), only cadre, educational qualification and training received retained in the final model which together explained 35 per cent variation in the dependent variable. All the three gave almost equal level of contribution (10%) for the dependent variable.

From the above finding, it could be noted that most of the parameters of functional linkage between research and extension under the study were least affected by the personal, psychological and organization variables. Few variables, *viz.*, cadre, educational qualification, training received and attitude were found to have exerting some influence on the linkage. As discussed earlier, the extent of functional linkage between research and extension was quite disheartening under all the systems. Further, the antecedent variables spoke little in the variation of the dependent variable. This certainly leads to conclusion that there could be some other reasons responsible for the existing strength of interaction between research and extension personnel. However, the need for involving lower cadre personnel who also possess lower educational level in the reciprocal interaction could not be ruled out.

Table 8.38 : Results of stepwise regression analysis (backward elimination method) of independent variables with the extent of overall functional linkage between research and extension personnel

Sl. No.	*Independent Variables*	*b-Values*	*t-Values*	*Relative Contribution (%)*
1.	Cadre (X_2)	7.0663	2.7448*	12.60
2.	Educational qualification (X_4)	-4.0841	2.3786*	12.00
3.	Training experience (X_6)	2.4022	2.4711*	11.40

* = Significant at 1 per cent level of significance.

R^2 Value = 0.35; *F* Value = 7.89* with 3 and 75 d.f.

8.3.2 Variables Affecting the Strength of Functional Linkage between Extension Personnel and Dairy Farmers

8.3.2.1 *Variables of Extension Personnel Affecting Their Extent of Functional Linkage with the Sampled Dairy Farmers*

In this sub-head, the selected background variables of extension personnel, *viz.*, personal, psychological and organizational variables were subjected for relational analysis as well as cause and effect determination. Apart from multiple regression, data were also subjected to the step wise regression analysis in order to find the most important determinants to the dependent variable. The results so obtained are discussed below :

From Table 8.39, it could be noted that the variables like attitude, job satisfaction, perception of management, external environment and organizational climate were positively and significantly associated whereas age, professional experience and family background of the EP had negative association with their extent of communication linkage (ECL) with dairy farmers. On multiple regression analysis, age, attitude, family background and morale of the EP and organizational climate were found to have significant effect on the dependent variable. All the selected independent variables could predict 49 per cent in the variation of dependent variable (Table 8.40). When data were subjected for stepwise regression analysis, in final model, three variables, *viz.*, attitude and organizational climate retained which made joint contribution of 32 per cent in the variation of extent of communication linkage of EP with dairy farmers. With respect to extent of linkage in planning and decision making (ELPDM), variables like family background of EP and the external environment had significant (1% level of significance) association with ELPDM (Table 8.39). On multiple regression analysis, three variables like professional experience, family background and external environment were found to have significant effect on the dependent variable, ELPDM (Table 8.40), jointly, all the 15 variables predicted 42 per cent in the variation of ELPDM. On stepwise regression analysis, in final model, two variables,

Table 8.39 : Zero order correlation coefficient of selected variables of extension personnel with their extent of functional linkages with farmers.

S. No.	Variables	Parameters of Functional Linkage					
		Communi-cation	*Planning & Decision Making*	*Implemen-tation & Evaluation*	*Supply and Services*	*Training*	*Overall*
1.	Cadre (X_2)	0.1192	-0.1275	0.0266	0.4238*	-0.0773	0.1062
2.	Age (X_3)	-0.3271**	-0.0445	-0.0988	-0.2306	0.0161	-0.1711
3.	Education (X_4)	-0.0895	0.1785	0.0275	-0.5365*	-0.0304	-0.1213
4.	Professional experience (X_5)	-0.3239**	-0.1667	-0.0744	-0.1861	-0.0779	-0.2028
5.	Training (X_6)	0.0163	0.1494	0.1556	0.2003	0.1872	0.1996
6.	Attitude (X_7)	0.2677**	-0.0196	-0.0511	0.1588	0.2659**	0.1370
7.	Family backround (X_8)	-0.0499	-0.2869**	-0.1327	-0.1067	-0.1432	-0.1900
8.	Achievement motivation (X_9)	0.1917	0.1469	0.0337	0.0155	0.1375	0.1060
9.	Value orientation (X_{10})	-0.0789	0.0270	-0.2152	-0.2443	-0.2477	-0.2081
10.	Job satisfaction (X_{11})	0.2589**	0.1668	0.1654	0.0885	0.2594**	0.2264
11.	Morale (X_{12})	-0.1490	-0.1769	-0.0487	0.1568	-0.2395	-0.0899
12.	Perception of management (X_{13})	0.2587**	-0.0420	-0.0239	0.1177	0.0910	0.0819
13.	External environment (X_{14})	0.2334	-0.0725	-0.1169	0.2048	0.3483*	0.1369
14.	Organizational climate (X_{15})	0.4589*	0.3124**	0.2921**	0.3552*	0.5524*	0.4754*
15.	Knowledge gap (X_{16})	0.0527	-0.0364	-0.0303	-0.2907**	-0.0155	0.0577

* = Significant at 1 per cent level of significance; ** = Significant at 5 per cent level of significance.

Table 8.40 : Multiple regression coefficient of the selected independent variables of extension personnel with their extent of functional linkage with farmers.

S. No.	*Variables*	*Parameters of Functional Linkage*											
		Communication		*Planning & Decision*		*Implementation & Making*		*Supply and Evaluation*		*Training Services*		*Overall*	
		b	*t*	*b*	*t*	*b*	*t*	*b*	*t*	*b*	*t*	*b*	*t*
1.	Cadre (X_2)	-7.1371	1.0432	-3.3476	0.3285	-1.5062	0.1714	9.6755	1.0139	-0.0333	0.0043	0.0694	0.0098
2.	Age (X_3)	-0.8204	1.2044	0.3171	0.3125	-0.8284	0.9466	-0.5938	0.6251	0.0726	0.0945	-0.4699	0.6680
3.	Education (X_4)	3.3606	1.0027	-2.6713	0.5350	1.3482	0.3131	-11.4947	2.4590**	-2.8331	0.7501	-1.8293	0.5286
4.	Professional experience (X_5)	0.0645	0.0916	-2.1360	2.0365	-0.1029	0.1138	-1.2598	1.2829	-0.7665	0.9661	-0.7921	1.0896
5.	Training (X_6)	0.2709	0.2249	2.3010	1.2820	2.0040	1.2947	4.0179	2.3910**	1.1680	0.8602	2.1291	1.7113**
6.	Attitude (X_7)	1.0650	1.6150	-0.7731	0.7824	-0.6744	0.7960	0.1448	0.1574	0.5786	0.7785	0.0402	0.0590
7.	Family background (X_8)	-1.2830	1.4195	-2.5390	1.8857**	-1.2644	1.0889	-1.5633	1.2401	-0.1977	0.1941	-1.3653	1.4629
8.	Achievement motivation (X_9)	0.3011	0.2797	-0.0618	0.0385	-0.8161	0.5899	-0.0578	0.0385	-0.3094	0.2550	-0.3274	0.2945
9.	Value orientation (X_{10})	-0.9177	0.9289	-1.7460	1.1863	-2.4375	1.9203**	-15382	1.1163	-2.2853	2.0526**	-1.9283	1.8902**
10.	Job satisfaction (X_{11})	0.1428	0.4855	0.6872	1.5679	0.5524	1.4613	0.6264	1.5264	0.4694	1.4158	0.5288	1.7405**

(Contd...)

S. No.	Variables	Parameters of Functional Linkage											
		Communi-cation		Planning & Decision		Implemen-tation & Making		Supply and Evaluation		Training Services		Overall	
		b	t	b	t	b	t	b	t	b	t	b	t
11.	Morale (X_{12})	0.7017	1.7196**	0.2084	0.3429	0.0959	0.1828	1.3272	2.3671**	-0.2445	0.5317	0.4077	0.9676
12.	Perception of management (X_{13})	0.4450	1.0602	-0.1402	0.2242	-0.0457	0.0848	0.1967	0.3360	-0.6059	1.2798	-0.0344	0.0794
13.	External environment (X_{14})	0.4239	0.2981	-5.0016	2.1184**	-3.5756	1.9571**	1.1008	0.5550	1.2199	0.7612	-1.2572	0.8562
14.	Organizational climate (X_{15})	0.4603	2.0969**	0.4353	0.3270	0.4745	1.6825**	0.2958	0.9661	0.5986	2.4197**	0.4381	1.9326**
15.	Knowledge gap (X_{16})	0.0131	0.0391	-0.3273	0.5004	0.0076	0.0176	0.5648	1.2054	-0.2070	0.5467	0.0294	0.0847
	R^2 values (F value at 15 and 310 d.f.)	**0.49 (1.60^{NS})**		**0.42 (1.67^{NS})**		**0.38 (1.15^{NS})**		**0.64 (3.65*)**		**0.55 (2.21**)**		**0.50 (1.61^{NS})**	

* = Significant at 1 per cent level of significance; ** = Significant at 5 per cent level of significance;
NS = Non-significant.

namely, professional experience and family background of EP were retained which made 21 per cent contribution in the variation of ELPDM.

Extension personnel's' extent of linkage in implementation and evaluation (ELIE) of field activities with dairy farmers significantly co-varied with their value orientation and organizational climate (Table 8.39). On multiple regression analysis (Table 8.40), besides above two variables, external environment of the extension department was found to have significance influence on the dependent variables. Jointly, all the 15 variables contributed 38 per cent (Table 8.42). On stepwise regression analysis, the above three variables retained in the final model which jointly contributed to the extent of 20 per cent only (Table 8.43).

With respect to extent of linkage in supply and services (ELSS), four variables of EP, *viz.*, age, education, achievement motivation and knowledge gap were negatively and significantly associated, and training, external environment and organizational climate were positively and significantly associated with their ELSS (Table 8.39). On multiple regression analysis, it was found that the variables like education, training and morale of the EP were having significant influence on the dependent variable ELSS. All the 15 variables contributed as high as 64 per cent in the variation of ELSS (Table 8.40). On stepwise regression analysis, in the final model, five variables namely education, professional experience, training, job satisfaction and morale retained which jointly contributed 52 per cent in the variation of ELSS (Table 8.44).

Similarly, when the score of extent of functional linkage in training (EFLT) of EP were correlated with their selected variables, it was found that attitude, job satisfaction, external environment, morale and organization climate were significantly associated with EFLT (Table 8.39). On multiple regression analysis, value orientation and organizational climate were found to have significant affect on the EFLT. Altogether 15 variables predicted 55 per cent in the variation of EFLT (Table 8.40). In final model, apart from above two, job satisfaction also appeared and jointly these three made 45 contribution among which organization climate had highest contribution (20.61%) (Table 8.45).

Besides the above five parameters of functional linkage between EP and DFs, overall extent of functional linkage (OEFL) of EP was also studied in relation to the selected variables of EP. On correlation analysis, it was found that professional experience, training, value orientation, job satisfaction and organizational climate were significantly associated with OEFL (Table 8.39). On multiple regression analysis, except training, above four variables had significance affect on OEFL. Altogether 15 variables contributed 50 per cent in the variation of OEFL (Table 8.40). On stepwise regression analysis, in final model, three variables namely training, value orientation and organizational climate appeared. Except training remaining two variables significantly affected the OEFL. Jointly, these three variables made 30 per cent contribution in the OEFL. The highest share (18.80%) was of organizational climate (Table 8.46).

From the findings, it could be derived that for strengthening the functional linkage of EP with their clients, the organizational climate needs to be improved on all of its dimensions. Extension personnel require advance and refresher training on subject matter as well as the extension methodology in order to get oriented towards modern value system and thus, ultimately, improving their linkage strength with the dairy farmers.

Table 8.41 : Results of stepwise regression analysis (backward elimination method) of independent variables of extension personnel with their extent of communication linkage with farmers.

Sl. No.	*Independent Variables*	*b-Values*	*t-Values*	*Relative Contribution (%)*
1.	Age (X_3)	-0.5657	2.2668**	10.65
2.	Attitude (X_7)	0.6708	1.3969^{NS}	9.35
3.	Organizational climate (X_{15})	0.4403	2.6096**	12.00

* = Significant at 1 per cent level of significance;

** = Significant at 5 per cent level of significance;

NS = Non significant. R^2 Value = 0.32; *F* Value = 6.44* with 3 and 43 d.f.

Table 8.42 : Results of stepwise regression analysis (backward elimination method) of independent variables of extension personnel with their extent of functional linkage in planning and decision making with farmers.

Sl. No.	*Independent Variables*	*b-Values*	*t-Values*	*Relative Contribution (%)*
1.	Professional experience (X_5)	-1.0973	2.2746**	12.85
2.	Family background (X_8)	-2.1363	1.8074**	8.15

* = Significant at 1 per cent level of significance;

** = Significant at 5 per cent level of significance.

R^2 Value = 0.21; *F* Value = 3.34* with 2 and 44 d.f.

Table 8.43 : Results of stepwise regression analysis (backward elimination method) of independent variables of extension personnel with their extent of functional linkage in implementation and evaluation with farmers.

Sl. No.	*Independent Variables*	*b-Values*	*t-Values*	*Relative Contribution (%)*
1.	Value orientation (X_{10})	-1.9063	1.9386**	6.45
2.	External environment (X_{14})	-2.6166	1.7458**	5.55
3.	Organizational climate (X_{15})	0.5085	2.4745**	9.00

* = Significant at 1 per cent level of significance;

** = Significant at 5 per cent level of significance.

R^2 Value = 0.20; *F* Value = 3.36* with 3 and 43 d.f.

Table 8.44 : Results of stepwise regression analysis (backward elimination method) of independent variables of extension personnel with their extent of functional linkage in supply and services with farmers.

Sl. No.	*Independent Variables*	*b-Values*	*t-Values*	*Relative Contribution (%)*
1.	Educational background (X_4)	-17.8105	4.8601*	14.15
2.	Professional experience (X_5)	-2.0965	3.5816*	10.25
3.	Training (X_6)	3.3946	2.1858**	9.30
4.	Job satisfaction (X_{11})	0.7030	2.0514**	8.70
5.	Morale (X_{12})	1.4328	3.2824*	9.60

* = Significant at 1 per cent level of significance;

** = Significant at 5 per cent level of significance.

R^2 Value = 0.52; *F* Value = 7.10* with 5 and 41 d.f.

Table 8.45 : Results of stepwise regression analysis (backward elimination method) of independent variables of extension personnel with their extent of functional linkage in training with farmers.

Sl. No.	*Independent Variables*	*b-Values*	*t-Values*	*Relative Contribution (%)*
1.	Value orientation (X_{10})	-2.0971	2.4657**	18.39
2.	Job satisfaction (X_{11})	0.3991	1.6822**	6.00
3.	Organizational climate (X_{15})	0.7760	4.2722*	20.61

* = Significant at 1 per cent level of significance;
** = Significant at 5 per cent level of significance.
R^2 Value = 0.45; F Value = 8.04* with 3 and 43 d.f.

Table 8.46 : Results of stepwise regression analysis (backward elimination method) of independent variables of extension personnel with their extent of overall functional linkage with farmers.

Sl. No.	*Independent Variables*	*b-Values*	*t-Values*	*Relative Contribution (%)*
1.	Training (X_6)	1.1962	1.1777^{NS}	7.20
2.	Value orientation (X_{10})	-1.3587	1.7193**	4.00
3.	Organizational climate (X_{15})	0.5882	3.5332*	18.80

* = Significant at 1 per cent level of significance;
** = Significant at 5 per cent level of significance; NS = Non-significant.
R^2 Value = 0.30; F Value = 5.96* with 3 and 43 d.f.

8.3.2.2 Variables of the Sampled Dairy Farmers Affecting Their Extent of Functional Linkage with Extension Personnel

The selected variables (21) of the sampled dairy farmers have been subjectively described in the following subheads as their profile :

8.3.2.2.1 Socio Personal Variables

(a) *Age :* From Table 8.47, it is evident that majority (65.56%) of the dairy farmers belonged to middle age group (between 29 to 53 years). The average age was worked out as 41 years and the age of sampled dairy farmers ranged from 22 to 75 years.

(b) ***Education :*** From the same table, it could be observed that majority of the dairy farmers (about 49%) were matriculate. However, a considerable percentage (30%) of them were found illiterate.

(c) ***Family Education Status :*** With respect to family education status (FES), about 71 per cent of the dairy farmers had medium level of FES, followed by about 19 and 10 per cent of them having high and low level of FES, respectively (Table 8.47).

(d) ***Family Type and Family Size :*** Regarding family type, about 51 per cent of the dairy farmers belonged to nuclear families. Marginally lesser percentage of dairy farmers (49%) were having joint families. Family size, however, showed slight variation and 57 per cent of the dairy farmers were found to have medium size of family (5 to 12 members), followed by 23.33 per cent and 19 per cent dairy farmers with large (more than 12 members) and small (upto 4 members) family size (Table 8.47).

(e) ***Occupation :*** From Table 8.47, it could be observed that for majority of the dairy farmers (73%), agriculture was the primary occupation. Only 6 per cent of them opted for dairying as primary occupation. Compared to dairying, slightly more percentage of the sampled farmers were found to have primary occupation as business (7%), service (8%) and contract labour (8%). With respect to secondary occupation, dairying was most choiced enterprise by about 63 per cent of the sampled farmers. Secondary occupation also showed a little variation. This leads to conclusion that there was considerable occupational diversity in the areas studied.

(f) ***Caste :*** Majority of the sampled farmers (about 59%) belonged to upper caste, followed by backward caste (31%) and scheduled caste (10%) by their social affiliation.

(g) ***Land Holding :*** A perusal of Table 8.47 shows that majority (34.44%) of the sampled farmers had medium

Table 8.47 : Profile of the dairy farmers sampled from selected extension systems.

Sl. No.	Variables	Categories	Scores	Extension Systems NDRI, Karnal (n=60)	KGK, Karnal (n=60)	SDAH, Karnal (n=60)	Pooled (n=180)	Mean	S.D.	Range
1	2	3	4	5	6	7	8	9	10	11
1.	Age	Young	Upto 28	7(11.67)	8(13.33)	14(23.33)	29(16.11)	41.10	12.32	22-75
		Middle	29 to 53	37(61.67)	43(71.67)	38(63.33)	118(65.56)			
		Old	More than 53	16(26.66)	9(15.00)	8(13.33)	33(18.33)			
2.	Education	Illiterate	0	20(33.33)	21(35.00)	13(21.67)	54(30.00)			0-17
		Middle	5	10(16.67)	10(16.67)	4(6.67)	24(13.33)			
		Matriculate	10	25(41.67)	27(45.00)	36(60.00)	88(48.88)			
		Intermediate	12	5(8.33)	–	4(6.67)	9(5.00)			
		Graduate	15	–	2(3.33)	2(3.33)	4(2.22)			
		Post-graduate	17	–	–	1(1.67)	1(1.67)			
3.	Family education status	Low	Less than 1.90	6(10.00)	5(8.33)	8(13.33)	19(10.59)	4.19	2.29	0.00-15.00
		Medium	1.90 to 6.48	47(78.33)	50(83.33)	30(50.00)	127(70.55)			
		High	More than 6.48	7(11.67)	5(8.33)	22(36.67)	34(18.89)			
4.	Family type	Nuclear	–	22(36.67)	29(48.33)	40(66.67)	91(50.56)			
		Joint	–	38(63.33)	31(51.67)	20(33.33)	89(49.44)			

(Contd...)

1	2	3	4	5	6	7	8	9	10	11
5.	Family size	Small	Upto 4	6(10.00)	9(15.00)	20(33.33)	35(19.44)	7.88	4.49	3-24
		Medium	5 to 12	37(61.67)	38(63.33)	38(63.33)	103(57.23)			
		Large	More than 12	17(28.33)	13(21.67)	2(3.33)	42(23.33)			
6.	Primary occupation	Agriculture	–	45(75.00)	50(83.33)	37(61.67)	132(73.33)			
		Dairying	–	2(3.33)	2(3.33)	7(11.67)	11(6.11)			
		Business	–	3(5.00)	6(10.00)	4(6.67)	13(7.22)			
		Contract labour	–	7(11.67)	1(1.67)	7(11.67)	15(8.33)			
		Service	–	4(6.67)	3(5.00)	8(13.33)	15(8.33)			
		Others	–	2(2.33)	–	–	2(1.11)			
7.	Secondary occupation	Agriculture	–	4(6.67)	5(8.33)	2(3.33)	11(18.33)			
		Dairying	–	39(65.00)	42(70.00)	32(53.33)	113(62.77)			
		Business	–	4(6.67)	7(11.67)	9(15.00)	29(11.11)			
		Contract labour	–	–	1(1.67)	3(5.00)	4(2.22)			
		Service	–	5(8.33)	1(1.67)	4(6.67)	10(5.55)			
		No secondary occupation	–	11(18.33)	6(10.00)	12(20.00)	29(16.11)			
8.	Caste	Upper caste	–	32(53.33)	40(66.67)	34(56.67)	106(58.89)			
		Backward caste	–	20(33.33)	15(25.00)	21(35.00)	56(31.11)			
		Scheduled caste	–	8(13.33)	5(8.33)	5(8.33)	18(10.00)			

(Contd...)

1	2	3	4	5	6	7	8	9	10	11
9.	Land holding (acres)	Landless	No land	9(15.00)	2(3.33)	14(23.33)	25(13.88)	7.50	6.40	0-40
		Marginal	Upto 2.50	8(13.33)	13(21.67)	8(13.33)	29(16.11)			
		Small	2.51 to 5.00	14(23.33)	15(25.00)	10(16.67)	39(21.66)			
		Medium	5.01 to 10.00	24(40.00)	20(33.33)	18(30.00)	62(34.44)			
		Large	Above 10.00	5(8.33)	10(16.67)	10(16.67)	25(13.88)			
10.	Herd size	Small	Upto 4	9(15.00)	7(11.67)	15(25.00)	31(17.22)	8.59	4.70	2-30
		Medium	5 to 13	42(70.00)	41(68.33)	35(58.33)	118(65.56)			
		Large	More than 13	9(15.00)	12(20.00)	10(16.67)	31(17.22)			
11.	Milk production (litres)	Low	Less than 6.70	13(21.67)	7(11.67)	12(20.00)	32(17.78)	14.96	8.26	3-90
		Medium	6.70 to 23.22	33(55.00)	48(80.00)	40(66.67)	121(67.22)			
		High	More than 23.22	14(23.33)	5(8.33)	8(13.33)	27(15.00)			
12.	Milk consumption (litres)	No consumption	0	8(13.33)	2(3.33)	4(6.66)	12(7.78)	8.80	4.81	0-30
		Low	Less than 4.08	3(5.00)	6(10.00)	13(21.67)	22(12.22)			
		Medium	4.08 to 13.71	37(61.67)	45(75.00)	34(56.67)	116(64.44)			
		High	More than 13.71	12(20.00)	7(11.67)	9(15.00)	28(15.56)			
13.	Milk sale (litres)	No sale	0	20(33.33)	21(35.00)	18(30.00)	59(32.78)	6.07	4.78	0-50
		Low	Less than 1.29	2(3.33)	1(1.67)	2(3.33)	5(2.78)			
		Medium	1.29 to 10.85	23(38.33)	33(55.00)	33(55.00)	89(49.44)			
		High	More than 10.85	15(25.00)	5(8.33)	7(11.67)	27(15.00)			

(Contd...)

1	2	3	4	5	6	7	8	9	10	11
14.	Social participation	No participation	0	20(33.33)	16(26.67)	24(40.00)	60(33.33)	1.91	0.81	0-5
		Low	Less than 1.10	10(16.67)	17(28.33)	15(25.00)	42(23.33)			
		Medium	1.10 to 2.72	17(28.33)	22(36.67)	11(18.33)	50(27.77)			
		High	More than 2.72	13(21.67)	5(8.33)	10(16.67)	28(15.56)			
15.	Extension contact	Low	Less than 7.29	8(13.33)	7(11.67)	2(3.33)	17(9.44)	14.93	7.64	4-40
		Medium	7.29 to 22.57	47(78.33)	48(70.00)	34(56.67)	129(71.67)			
		High	More than 22.57	5(8.33)	5(8.33)	24(40.00)	34(18.89)			
16.	Mass media exposure	No exposure	0	16(26.67)	27(45.00)	5(8.33)	48(26.67)	4.84	2.04	0-12
		Low	Less than 2.80	10(16.67)	13(21.67)	3(5.00)	26(14.44)			
		Average	2.80 to 6.88	29(48.33)	15(25.00)	35(58.33)	79(43.89)			
		High	More than 6.88	5(8.33)	5(8.33)	17(28.33)	27(15.00)			
17.	Risk preference	Low	Upto 23.33	16(26.67)	12(20.00)	8(13.33)	36(20.00)	27.16	3.82	20-36
		Medium	23.34 to 30.98	32(53.33)	42(70.00)	38(63.33)	112(62.22)			
		High	More than 30.98	12(20.00)	6(10.00)	14(23.33)	32(17.78)			
18.	Cosmopoliteness	Low	Less than 16.02	10(16.67)	9(15.00)	8(13.33)	27(15.00)			
		Medium	16.02 to 20.16	42(70.00)	47(78.33)	41(68.33)	130(72.22)			
		High	More than 20.16	8(13.33)	4(6.67)	11(18.33)	23(12.78)			
19.	Source perception	Self	–	55(91.67)	9(15.00)	50(83.33)	114(63.33)			
		Relatives	–	–	–	–	–			

(Contd...)

1	2	3	4	5	6	7	8	9	10	11
		Friends	–	–	10(16.67)	–	10(5.56)			
		Extension staff	–	5(8.33)	33(55.00)	10(16.67)	48(26.67)			
		No perception	–	–	8(13.33)	–	8(4.44)			
20.	Perception of management	No perception	0	24(40.00)	30(50.00)	23(38.33)	77(42.78)	51.68	2.94	0-58
		Low (Poor)	Upto 48.73	8(13.33)	10(5.00)	9(15.00)	27(15.00)			
		Medium (Average)	48.74 to 54.62	24(40.00)	14(23.33)	24(40.00)	62(34.44)			
		High (Good)	Above 54.62	4(6.67)	6(10.00)	4(6.67)	14(7.77)			
21.	Knowledge about the department	No knowledge	0	–	19(31.67)	–	19(10.56)	55.52	15.07	0-86.67
		Poor	Upto 40.52	5(8.33)	18(30.00)	12(20.00)	35(19.44)			
		Average	40.53 to 70.59	44(73.33)	21(35.00)	32(53.33)	37(53.89)			
		Good	Above 70.59	11(18.33)	2(3.33)	16(26.67)	29(16.11)			

holding of land (5 to 10 acres), followed by small (22%), marginal (16%) and large (14%) land holders. A considerable percentage (about 14%) of the farmers were also found landless. The findings reflect that in the study area, the distribution of land holding was not skewed.

(h) ***Herd Size :*** Majority of the dairy farmers (66%) in sample possessed medium herd size (5 to 13 animals), followed by equal number of them (17%) having small (upto 4 animals) and large (more than 12 animals) herd.

(i) ***Milk Production, Milk Consumption and Milk Sale :*** From the same table, it could be observed that the average level of milk production, consumption and sale were about 15, 9 and 6 kg, respectively. Majority of dairy farmers belonged to a medium category with respect to milk production (67%), milk consumption (64%) and milk sale (49%). A substantial percentage (33%) of the dairy farmers were found not selling the milk. This reflect that home consumption level of milk and milk products in the study area was considerably high.

(j) ***Social Participation :*** Majority (33%) of the dairy farmers in the study area were found to have no social participation, followed by about 28, 23 and 15 per cent of them having medium, low and high level of social participation, respectively.

(k) ***Extension Contact :*** From Table 8.47, it is evident that most of the dairy farmers (72%) had medium level of extension contact, followed by 19 and 9 per cent of them having high and low extent of extension contact, respectively.

(l) ***Mass Media Exposure :*** With respect to mass media exposure of the dairy farmers, it was found that majority (44%) of them had average exposure to mass media, followed by high (15%) and low (14%) level of such exposure. It was found that about 27 per cent of the farmers did not have exposure to any type of mass media.

Table 8.48 : Frequency distribution of the sampled dairy farmers based on their knowledge about the activities of the extension organizations/departments

Extension Organizations/ Department	*Categories of of the Respondents*	*Aims of the Organization/Department (Each digit indicates number of aims)*				*Time of Inception*		*Facilities offered by the Organizations/Department (Each digit indicates number of facilities)*				
		0	***1***	***2***	***3***	*W* ***1***	*R*	***0***	***1***	***2***	***3***	***4***
NDRI	Small (n=21)	0 (0.00)	7 (33.33)	12 (57.14)	2 (9.53)	5 (23.81)	16 (76.19)	0 (0.00)	3 (14.28)	6 (28.59)	9 (42.85)	3 (14.28)
	Medium (n=27)	0 (0.00)	6 (22.22)	21 (77.78)	0 (0.00)	0 (0.00)	27 (100.00)	0 (0.00)	0 (0.00)	10 (37.04)	16 (59.26)	1 (3.70)
	Large (n=12)	0 (0.00)	0 (0.00)	10 (83.33)	2 (16.67)	2 (16.67)	10 (83.33)	0 (0.00)	0 (0.00)	0 (0.00)	4 (33.33)	8 (66.67)
	Pooled (n=60)	0 (0.00)	13 (21.67)	43 (71.67)	4 (6.66)	7 (11.67)	53 (88.33)	0 (0.00)	3 (5.00)	16 (26.67)	29 (48.33)	12 (20.00)
HAU	Small (n=22)	12 (54.55)	4 (18.18)	6 (27.27)	0 (0.00)	18 (81.82)	4 (18.18)	11 (50.00)	6 (27.27)	4 (18.18)	1 (4.55)	0 (0.00)
	Medium (n=25)	4 (16.00)	14 (58.00)	7 (28.00)	0 (0.00)	15 (60.00)	10 (40.00)	3 (12.00)	3 (12.00)	11 (44.00)	8 (32.00)	0 (0.00)
	Large (n=13)	6 (46.15)	2 (15.38)	5 (38.47)	0 (0.00)	11 (84.62)	2 (15.38)	5 (38.46)	4 (30.76)	3 (23.08)	1 (7.70)	0 (0.00)
	Pooled (n=60)	22	20	18	0 (0.00)	34	16	19	13	18	10	0 (0.00)

(Contd...)

		0	*1*	2	*3*	*1*		*0*	*1*	*2*	*3*	*4*
SDAH	Small (n=20)	0 (0.00)	10 (50.00)	10 (50.00)	0 (0.00)	4 (20.00)	16 (80.00)	0 (0.00)	0 (0.00)	13 (65.00)	6 (30.00)	1 (5.00)
	Medium (n=25)	0 (0.00)	11 (44.00)	7 (28.00)	7 (28.00)	4 (16.00)	21 (84.00)	0 (0.00)	9 (36.00)	2 (8.00)	10 (40.00)	4 (16.00)
	Large (n=15)	0 (0.00)	3 (20.00)	8 (53.33)	4 (26.67)	1 (6.67)	14 (93.33)	0 (0.00)	1 (6.67)	3 (20.00)	3 (20.00)	8 (53.33)
	Pooled (n=60)	0 (0.00)	24 (40.00)	25 (41.67)	11 (18.33)	9 (11.00)	51 (89.00)	0 (0.00)	10 (16.67)	18 (30.00)	19 (31.67)	13 (21.67)
Overall	Small (n=63)	12 (19.05)	21 (33.33)	28 (44.44)	2 (3.18)	27 (42.86)	36 (57.14)	11 (17.46)	9 (14.29)	23 (36.51)	16 (25.39)	4 (6.35)
	Medium (n=77)	4 (15.19)	31 (40.26)	35 (45.45)	7 (9.09)	19 (24.68)	58 (75.32)	3 (3.89)	12 (15.58)	23 (29.87)	34 (44.16)	5 (6.50)
	Large (n=40)	6 (15.00)	5 (12.50)	23 (57.50)	6 (15.00)	14 (35.00)	26 (65.50)	5 (12.50)	5 (12.50)	6 (15.00)	8 (20.00)	16 (40.00)
	Pooled (n=180)	22 (12.22)	57 (31.67)	86 (47.78)	15 (8.33)	60 (33.33)	120 (66.67)	19 (10.56)	26 (14.44)	52 (28.89)	58 (32.22)	25 (13.89)

(Contd...)

Extension Organizations/ Department	*Categories of the Respondents*	*Service Offered to*		*Employment Opportunity*		*Fodder Development*		*Organizing*		*Computing Campaign*		*Conducting Trials and Demonstration*	
		W	R	W	R	W	R	W	R	W	R	W	R
		1	2	1	2	1	2	1	2	1	2	1	2
NDRI	Small (n=21)	0 (0.00)	21 (100.00)	18 (85.71)	3 (14.29)	15 (71.43)	6 (28.57)	3 (14.28)	18 (85.72)	14 (66.67)	7 (33.33)	18 (85.71)	3 (14.29)
	Medium (n=27)	0 (0.00)	26 (96.30)	18 (66.67)	9 (33.33)	20 (74.07)	7 (25.93)	0 (0.00)	27 (100.00)	5 (18.52)	22 (81.48)	21 (77.78)	6 (22.22)
	Large (n=12)	0 (0.00)	12 (100.00)	7 (58.33)	5 (41.67)	5 (41.67)	7 (58.33)	0 (0.00)	12 (100.00)	5 (41.67)	7 (58.33)	9 (75.00)	3 (25.00)
	Pooled (n=60)	1 (1.67)	59 (98.33)	43 (71.67)	17 (28.33)	40 (66.67)	20 (33.33)	3 (5.00)	57 (95.00)	24 (40.00)	36 (60.00)	48 (80.00)	12 (20.00)
HAU	Small (n=22)	12 (54.55)	10 (45.45)	21 (95.45)	1 (4.45)	16 (72.73)	6 (27.27)	16 (72.73)	6 (27.27)	21 (95.45)	1 (4.55)	13 (59.10)	9 (40.90)
	Medium (n=25)	3 (12.00)	22 (88.00)	19 (76.00)	6 (24.00)	16 (64.00)	9 (36.00)	14 (56.00)	11 (44.00)	23 (92.00)	2 (8.00)	11 (44.00)	14 (56.00)
	Large (n=13)	6 (46.15)	7 (15.38)	13 (100.00)	0 (0.00)	5 (38.46)	8 (61.54)	8 (61.54)	5 (38.46)	11 (84.62)	2 (15.38)	8 (61.54)	5 (38.46)
	Pooled (n=60)	21	39	53	7	37	23	38	22	55	5	32	28

(Contd...)

		W	R	W	R	W	R	W	R	W	R	W	R
		1	2	1	2	1	2	1	2	1	2	1	2
SDAH	Small (n=20)	0 (0.00)	20 (100.00)	20 (100.00)	0 (0.00)	16 (80.00)	4 (20.00)	0 (0.00)	20 (100.00)	3 (15.00)	17 (85.00)	17 (85.00)	3 (15.00)
	Medium (n=25)	2 (8.00)	23 (92.00)	25 (100.00)	0 (0.00)	15 (60.00)	10 (40.00)	4 (16.00)	21 (84.00)	13 (52.00)	12 (48.00)	24 (96.00)	1 (4.00)
	Large (n=15)	0 (0.00)	15 (100.00)	15 (100.00)	0 (0.00)	9 (60.00)	6 (40.00)	1 (6.67)	14 (93.33)	0 (0.00)	15 (100.00)	10 (66.67)	5 (33.33)
	Pooled (n=60)	2 (3.33)	58 (96.67)	60 (100.00)	0 (0.00)	40 (66.67)	20 (33.33)	5 (8.33)	55 (26.67)	16 (73.33)	44 (85.00)	51 (85.00)	9 (15.00)
Overall	Small (n=63)	12 (19.05)	51 (80.95)	59 (93.65)	4 (6.35)	47 (74.60)	16 (25.40)	19 (30.16)	44 (69.84)	38 (60.32)	25 (39.68)	48 (76.19)	15 (23.81)
	Medium (n=77)	6 (7.79)	71 (92.21)	62 (80.52)	15 (19.48)	51 (66.23)	26 (33.77)	18 (23.38)	59 (76.62)	41 (53.25)	36 (46.75)	56 (72.73)	21 (27.27)
	Large (n=40)	6 (15.00)	34 (85.00)	35 (87.50)	5 (12.50)	19 (47.50)	21 (52.50)	9 (22.50)	31 (77.50)	16 (40.00)	24 (60.00)	27 (67.50)	13 (32.50)
	Pooled (n=180)	24 (13.33)	156 (86.67)	156 (86.67)	24 (13.33)	117 (65.00)	63 (35.00)	46 (25.56)	134 (74.44)	95 (52.78)	85 (47.22)	131 (72.78)	49 (27.22)

Figures in parentheses indicate percentage. W = Wrong; R = Right.

(m) ***Risk Preference :*** Risk preference also followed the trends of medium, low and high level by 62, 20 and 18 per cent of the sampled dairy farmers, respectively.

(n) ***Cosmopoliteness :*** Majority (72%) of the farmers were found to have medium level of cosmopoliteness. This was followed by 15 and 13 per cent of the farmers having low and high degree of cosmopoliteness.

(o) ***Source Perception :*** From Table 8.47, it could be seen that most of the dairy farmers (63%) had the perception about the extension departments and their activities by themselves. About 27 per cent of the farmers were found to have source perception through the staff of the department.

(p) ***Perception of Management :*** Regarding the farmers' perception of the management of the department, it was found that about 43 per cent of them did not have any idea about the management activities of the department. About 32 per cent of them were in medium level of perception of management, followed by 15 and 8 per cent in low and high perception categories.

(q) ***Knowledge About the Department :*** From Table 8.47, it is evident that most of the dairy farmers (54%) had medium level of knowledge about the department, followed by poor (19.44%) and good (16%) level of knowledge. It was found that about 11 per cent of them did not have any knowledge about the extension department of their area. For a comprehensive understanding of knowledge of dairy farmers about the department, findings are presented in Table 8.48.

8.3.2.2.2 Gap in the Adoption of the Scientific Dairy Farming Practices (SDFP)

A perusal of Table 8.49 reveals that the highest (about 36%) level of gap of adoption in the SDFP was noted among the farmers of adopted villages by SDAH extension system. This was followed by 33 per cent gap in the adoption of SDFP among the farmers from adopted villages of HAU. Comparatively, lowest gap (31.25) in the adoption of SDFP was found among the farmers of NDRI adopted villages. The probable reason behind better performance of NDRI extension system could be the existence of village level institutions, concentration of efforts, lower areas of coverage and hence better availability of the facilities. Further, it was also found that irrespective of categories and system, highest gap in the adoption of SDFP was in the area of fodder production (40%) and lowest in breeding practices (28%).

Table 8.49 : Average extent of gap in the adoption of recommended dairy farming practices among the farmers sampled from selected extension systems.

S. No.	Extension systems	Categories of Farmers	Sample Size	Areas of Improved Dairy Farming: Breeding	Feeding	Health Care	Management	Fodder Product	Overall
1.	NDRI	Small	21	33.45	33.33	36.31	29.76	56.35	37.84
		Medium	27	25.87	25.93	22.68	35.53	35.50	29.10
		Large	12	30.21	20.83	20.83	29.69	29.17	26.15
		Pooled	60	29.84	26.70	26.61	31.66	40.67	31.25
2.	HAU	Small	22	35.79	33.18	38.98	33.80	62.87	40.92
		Medium	25	20.01	25.60	31.50	24.47	53.99	31.11
		Large	13	24.04	20.00	27.88	25.96	24.61	24.43
		Pooled	60	26.61	26.26	32.79	28.08	47.16	32.79
3.	SDAH	Small	20	33.75	42.00	27.75	47.74	50.00	40.25
		Medium	25	31.43	34.83	44.83	45.36	33.12	37.91
		Large	15	20.83	26.00	22.50	42.10	12.69	24.82
		Pooled	60	28.63	34.28	31.69	45.07	31.94	36.45
4.	Overall	Small	63	34.33	36.17	34.35	37.10	56.40	39.67
		Medium	77	25.77	28.78	33.00	35.12	40.87	32.71
		Large	40	25.03	22.78	23.74	32.58	22.16	25.26
		Pooled	180	28.38	29.24	30.36	34.93	39.81	33.41

The extent of knowledge gap in the SDFP was also computed. From Table 8.50, it could be observed that the highest (about 29%) gap in the extent of knowledge about SDFP was in case of

HAU extension system, followed by the gap of about 27 per cent and 24 per cent in case of the dairy farmers taken from NDRI and SDAH extension systems, respectively.

Further, when knowledge adoption gap was subjected for statistical validation, it was found that the difference was significant in all the three selected systems. With respect to linkage, the significance of knowledge-adoption gap was duly recognised by Singh (1995). The significant knowledge-adoption gap in the current investigation speak adequately on weaker status of extension farmers linkage as well as linkages of other types.

Table 8.50 : Average extent of knowledge gap, adoption gap and knowledge adoption gap of the dairy farmers sampled from the selected extension systems.

Sl.No. Particulars	*Extension Systems*			
	NDRI (N=60)	*HAU (N=60)*	*SDAH (N=60)*	*Pooled (N=180)*
1. Knowledge gap	26.81	28.77	23.67	**26.41**
2. Adoption gap	31.25	32.79	36.45	**33.41**
3. Knowledge - Adoption gap	4.44 (4.73 at 58 d.f.)	4.02 (3.74 at 58 d.f.)	12.78 (9.22 at 58 d.f.)	**7.08 (9.77 at 158 d.f.)**

8.3.2.2.3 Correlation and Regression Analyses

The antecedent variables of dairy farmers have been studied and explained subjectively in the above subheads. In order to find the association as well as cause and effect relationship between these variables of dairy farmers with their extent of functional linkage with extension personnel, data were subjected for the statistics like zero order correlation, multiple regression and stepwise regression analyses. The findings have been presented and discussed in this subhead.

Findings in Table 8.51 reveal that except age, family type, family size and secondary occupation of dairy farmers, all 17 variables showed significant co-variation with the extent of communication linkage (ECL). On multiple regression analysis, three variables, *viz.*, social participation, extension contact and

Table 8.51 : Zero order correlation coefficient of selected variables of dairy farmers with their extent of functional linkage with the extension personnel.

Sl.No.	Variables	*Parametes of Functional Linkage*					
		Communi-cation (X_{22})	*Planning and Decision Making (X_{23})*	*Implemen-tation and Evaluation (X_{24})*	*Supply and Services (X_{25})*	*Training (X_{26})*	*Overall (X_{27})*
1.	Age (X_1)	-0.0642	-0.1500**	-0.1934*	-0.1351	-0.1245	-0.1686**
2.	Education (X_2)	0.2304*	0.1630**	0.2603*	0.2425*	0.1621**	0.2938*
3.	Family education status (X_3)	0.2485*	0.1942*	0.2727*	0.3016*	0.0243	0.2909*
4.	Family type (X_4)	0.0071	-0.0165	-0.1343	-0.0646	0.0728	-0.0692
5.	Family size (X_5)	0.0596	0.0130	-0.0990	-0.0779	0.0703	-0.0286
6.	Primary occupation (X_6)	-0.1755**	-0.2010*	-0.1640**	-0.0144	-0.1596**	-0.1533**
7.	Secondary occupation (X_7)	0.1109	0.1106	0.0147	0.0596	0.0273	0.0786
8.	Caste (X_8)	0.1674**	0.2663*	0.1303	0.0451	0.1143	0.1577**
9.	Land holding (X_9)	0.2380*	0.3264*	0.2536*	0.1617**	0.0265	0.2316*
10.	Herd size (X_{10})	0.2011*	0.2115*	0.2126*	0.0389	0.0508	0.1658**
11.	Milk production (X_{11})	0.3429*	0.1772**	0.2346*	0.2351*	0.0581	0.3064*
12.	Milk consumption (X_{12})	0.3025*	0.2409*	0.2071*	0.1081	0.0912	0.2367*
13.	Milk sale (X_{13})	0.2660*	0.0944	0.1810**	0.2293*	0.0246	0.2526*

(Contd...)

Sl.No. Variables	Parametes of Functional Linkage					
	Communication (X_{22})	Planning and Decision Making (X_{23})	Implementation and Evaluation (X_{24})	Supply and Services (X_{25})	Training (X_{26})	Overall (X_{27})
14. Social participation (X_{14})	0.2953*	0.2607*	0.1047	0.1720**	0.2303*	0.2308*
15. Extension contact (X_{15})	0.5184*	0.4336*	0.4353*	0.4673*	0.1824**	0.5524*
16. Mass media exposure (X_{16})	0.5278*	0.3439*	0.4548*	0.5278*	0.1453	0.5605*
17. Risk preference (X_{17})	0.3442*	0.1948*	0.3500*	0.2764*	0.1451	0.3402*
18. Localite cosmopoliteness (X_{18})	0.2939*	0.1312	0.2809*	0.2269*	0.1515	0.2737*
19. Source perception (X_{19})	0.6729*	0.1686**	0.4988*	0.6929*	0.1379	0.6701*
20. Perception of management (X_{20})	0.3688*	0.2456*	0.3983*	0.2817*	0.0798	0.3800*
21. Knowledge about the department (X_{21})	0.8402*	0.3465*	0.6825*	0.7259*	0.2936*	0.8230*

knowledge about the department were found to have significant affect on ECL. All the 21 variables jointly predicted 78 per cent in the variation of ECL (Table 8.52). On stepwise multiple regression analysis, the final model, three variables as above retained of which jointly predicted 76 per cent in the variation of ECL (Table 8.53). The maximum contribution of 30 per cent was by the farmers' knowledge about the department.

Similarly, with respect to farmers' extent of functional linkage in planning and decision making (EFLPDM) except the four variables, *viz.*, family type, family size, milk sale and localite-cosmopolite, remaining 17 variables had significant association with EFLPDM (Table 8.51). On multiple regression analysis, variables like age, land holding, social participation, extension contact, source perception and knowledge about the department were found having significant effect on EFLPDM. Jointly, all the 21 independent variables made 78 per cent contribution in the variation of dependent variable (Table 8.52). In the final model, after stepwise regression analysis, six variables namely age, milk sale, social participation, extension contact, source perception and farmers' knowledge about the department retained which jointly made 76 per cent contribution in the variation of dependent variable (Table 8.54).

Similarly, farmers' extent of functional linkage in implementation and evaluation (EFLIE) of field activities with EP, it was found that barring 5 variables namely, family type, family size, secondary occupation, caste and social participation, and rest of the 16 variables co-varied significantly with EFLIE (Table 8.51). On multiple regression analysis, 5 variables like family size, herd size, extension contact, perception of management and knowledge about department were affecting EFLIE significantly. Jointly, 21 independent variables made 60 per cent contribution (Table 8.52). On stepwise regression analysis apart from above 5 variables, milk production and milk consumption also appeared and together all seven variables made 57 per cent contribution in the variation of farmers' EFLIE.

With respect to farmers' extent of functional linkage in supply and services (EFLSS), variables like education, family

Table 8.52 : Zero order correlation coefficient of selected variables of dairy farmers with their extent of functional linkage with the extension personnel.

Sl.No.	Variables	Parametes of Functional Linkage											
		Communication (X_{22})		*Planning and Decision Making (X_{23})*		*Implementation and Evaluation (X_{24})*		*Supply and Services (X_{25})*		*Training (X_{26})*		*Overall (X_{27})*	
		b	*t*	*b*	*t*	*b*	*t*	*b*	*t*	*b*	*t*	*b*	*t*
1.	Age (X_1)	-0.0689	1.0528	-0.2399	2.3405*	-0.0788	0.9474	-0.1920	1.2286	-0.0922	2.3468*	-0.1084	2.0456**
2.	Education (X_2)	-0.2192	1.2684	-0.0690	0.2601	-0.1913	0.8836	-0.4677	1.1558	0.0563	0.5532	-0.0711	0.5178
3.	Family education status (X_3)	0.2461	0.8162	-0.1004	0.2087	0.0836	0.2141	1.4192	1.9350**	-0.2263	1.2273	0.2108	0.8471
4.	Family type (X_4)	-1.0967	0.7259	-1.4117	0.6094	-2.6468	1.4081	1.7475	0.4947	0.5066	0.5704	-1.2684	1.0586
5.	Family size (X_5)	0.0259	0.1163	-0.3552	1.0410	-0.4970	1.7956**	-0.3165	0.6085	0.1361	1.0404	-0.0428	0.2427
6.	Primary occupation (X_6)	-0.3413	0.6244	1.1349	1.3538	0.4215	0.6197	0.3889	0.3043	-0.4275	1.3303	-0.1276	0.2942
7.	Secondary occupation (X_7)	0.1835	0.3301	0.7335	0.8839	0.1747	0.2526	0.7666	0.5899	-0.1127	0.3449	0.1374	0.3116
8.	Caste (X_8)	-0.6631	0.6020	2.0580	1.2183	-1.0553	0.7700	-2.5397	0.9862	-0.0947	0.1463	-0.3186	0.3647
9.	Land holding (X_9)	-0.1402	1.2139	0.3155	1.7820**	0.1908	1.3279	-0.0375	0.1391	-0.1339	1.9734**	-0.0766	0.8370
10.	Herd size (X_{10})	-0.2478	1.1930	0.1536	0.4820	0.4707	1.8218**	-1.1962	2.4635**	-0.0245	0.2010	-0.2036	1.2370

(Contd...)

	b	*t*	*b*	*t*	*b*	*t*	*b*	*t*	*b*	*t*	*b*	*t*
11. Milk production (X_{11})	1.8429	1.1170	1.6002	0.6325	-0.9996	0.4870	4.2201	1.0941	0.6137	0.6328	1.8978	1.4505
12. Milk consumption (X_{12})	-1.5636	0.9657	-1.3230	0.5328	1.2121	0.6017	-3.8617	1.0201	-0.7925	0.8326	-1.6953	1.3202
13. Milk sale (X_{13})	-1.8051	1.0982	-1.8621	0.7388	0.7842	0.3835	-3.8648	1.0058	-0.6460	0.6626	-1.8795	1.4419
14. Social participation (X_{14})	1.7543	2.6108*	1.9652	1.9074**	-0.4492	0.5374	2.5802	1.6426	1.4883	3.7683*	0.9811	1.8412**
15. Extension contact (X_{15})	0.3711	3.4000*	0.5014	2.9963*	0.2406	1.7722**	0.5657	2.2172**	0.1238	1.9299**	0.3751	4.3335*
16. Mass media exposure (X_{16})	0.0776	0.2712	-0.2849	0.6491	0.0456	0.1281	1.0449	1.5613	-0.1636	0.9720	0.1487	0.6550
17. Risk preference (X_{17})	-0.0544	0.2503	0.0203	0.0609	0.1884	0.6971	-0.2892	0.5693	-0.0299	0.2342	-0.0776	0.4504
18. Localite cosmopoliteness (X_{18})	0.6078	1.5322	-0.8897	1.4627	-0.1044	0.2116	1.2891	1.3901	0.1959	0.8400	0.1468	0.4668
19. Source perception (X_{19})	0.9059	1.0177	-3.9348	2.8827*	-1.7508	1.5809	8.3180	3.9971*	-1.1896	2.2735**	0.4579	0.6487
20. Perception of management (X_{20})	-0.0194	0.7697	0.0215	0.5566	0.0737	2.3555*	-0.0971	1.6516**	-0.0224	1.5169	-0.0044	0.2201
21. Knowledge about the department (X_{21})	0.4347	8.9216*	0.2972	3.9785*	0.4054	6.6890*	0.4162	3.6537*	0.1127	3.9338*	0.3297	8.5325*
R^2 values (*F* value at 21 and 158 d.f.)	0.78 (26.99*)		0.38 (4.47*)		0.60 (11.03*)		0.68 (15.62*)		0.30 (2.78*)		0.78 (26.40*)	

* = Significant at 1 per cent level of significance; ** = Significant at 5 per cent level of significance.

education status, caste, milk production, milk sale, and the communication and psychological variables and source perception, perception of management and knowledge about department showed significant correlation with the EFLSS (Table 8.51). On multiple regression analysis, variables like family educational status, herd size, extension contact, source perception, perception of management and knowledge about department found to have significant influence on the dependent variable EFLSS. Altogether, 21 independent variables contributed to the extent of 68 per cent in the variation of EFLSS (Table 8.52). On stepwise regression analysis, besides above 5 variables, milk production also appeared in final model and all these 6 variables predicted to the extent of 65 per cent in the variation of EFLSS (Table 8.53).

Table 8.53 : Results of stepwise regression analysis (backward elimination method) of independent variables of dairy farmers with their extent of communication linkage with extension personnel.

Sl. No.	*Independent Variables*	*b-Values*	*t-Values*	*Relative Contribution (%)*
1.	Social participation (X_{14})	1.2267	2.428*	21.00
2.	Extension contact (X_{15})	0.3993	5.2364*	20.00
3.	Knowledge about the department (X_{21})	0.4771	18.6704*	30.00

* = Significant at 1 per cent level of significance;
** = Significant at 5 per cent level of significance.
R^2 Value = 0.76; F Value = 188.54* with 3 and 176 d.f.

Dairy farmers' extent of functional linkage in training (EFLT) with the EP was found significantly associated with their variables like education, primary occupation, social participation, extension contact and knowledge about the department (Table 8.51). Multiple regression analysis revealed that the variables like age, land holding, social participation, extension contact, source perception and knowledge about the department had significant influence on the dependent variables. All the 21 variables predicted only 30 per cent in the variation of EFLT (Table 8.52). On stepwise regression analysis except extension contact, all the above four variables retained and they jointly made 21 per cent contribution to the variation of EFLT (Table 8.57).

Table 8.54 : Results of stepwise regression analysis (backward elimination method) of independent variables of dairy farmers with their extent of functional linkage in planning and decision making with extension personnel.

Sl. No.	*Independent Variables*	*b-Values*	*t-Values*	*Relative Contribution (%)*
1.	Age (X_1)	-0.2280	3.0288*	13.16
2.	Milk Sale (X_{13})	-0.1697	2.0028**	8.24
3.	Social participation (X_{14})	2.4889	2.8873*	10.00
4.	Extension contact (X_{15})	0.5551	4.3443*	18.38
5.	Source perception (X_{19})	-3.5695	2.8489*	10.72
6.	Knowledge about the department (X_{21})	0.2623	3.9823*	16.50

* = Significant at 1 per cent level of significance;
** = Significant at 5 per cent level of significance.
R^2 Value = 0.76; F Value = 13.69* with 6 and 173 d.f.

Table 8.55 : Results of stepwise regression analysis (backward elimination method) of independent variables of dairy farmers with their extent of functional linkage in implementation and evaluation with extension personnel.

Sl. No.	*Independent Variables*	*b-Values*	*t-Values*	*Relative Contribution (%)*
1.	Family size (X_5)	-0.7748	4.0645*	6.23
2.	Herd size (X_{10})	0.5262	2.3406*	8.37
3.	Milk production (X_{11})	-0.2088	2.6153*	8.35
4.	Milk consumption (X_{12})	0.4165	2.0857**	6.20
5.	Extension contact (X_{15})	0.2809	2.7063*	8.66
6.	Perception of management (X_{20})	0.0722	2.5218*	8.44
7.	Knowledge about the department (X_{21})	0.3355	9.7294*	10.75

* = Significant at 1 per cent level of significance;
** = Significant at 5 per cent level of significance.
R^2 Value = 0.57; F Value = 32.63* with 7 and 172 d. f.

The extent of overall functional linkage (EOFL) of dairy farmers with EP was significantly correlated with all the variables except family type, family size and secondary occupation of the dairy farmers (Table 8.51). On multiple regression analysis, only 4 variables like age, social participation, extension contact and knowledge about the department were found to have significant affect on farmers' EOFL with EP. Jointly, all the 21 variables explained 78 per cent in the variation of dependent variables (Table 8.52). In the final model, social participation, in addition to the above four variables retained which predicted 77 per cent in the variation of EOFL of dairy farmers with EP.

Table 8.56 : Results of stepwise regression analysis (backward elimination method) of independent variables of dairy farmers with their extent of functional linkage in supply and services with extension personnel

Sl. No.	*Independent Variables*	*b-Values*	*t-Values*	*Relative Contribution (%)*
1.	Family education status (X_3)	1.6044	2.5855*	8.48
2.	Herd size (X_{10})	-1.2644	3.2266*	10.62
3.	Milk production (X_{11})	0.3366	2.2052**	7.52
4.	Extension contact (X_{15})	0.7932	3.6826*	11.27
5.	Source perception (X_{19})	8.0119	4.1075*	14.38
6.	Knowledge about the department (X_{21})	0.4207	4.0853*	11.73

* = Significant at 1 per cent level of significance;
** = Significant at 5 per cent level of significance.
R^2 Value = 0.65; *F* Value = 52.97* with 6 and 173 d.f.

8.3.3 *Variables Affecting the Strength of Functional Linkage Between Research Personnel and Dairy Farmers*

Under this subhead, the selected variables of research personnel were subjected to correlation and regression analyses with their extent of communication linkage (ECL) with the dairy farmers. Findings are contained in Table 8.59 and 8.60.

Table 8.57 : Results of stepwise regression analysis (backward elimination method) of independent variables of dairy farmers with their extent of functional linkage in training with extension personnel

Sl. No.	*Independent Variables*	*b-Values*	*t-Values*	*Relative Contribution (%)*
1.	Age (X_1)	-0.0957	3.3715*	5.29
2.	Land holding (X_9)	-0.1100	2.0868**	1.00
3.	Social participation (X_{14})	1.3344	3.9368*	5.81
4.	Source perception (X_{19})	-1.2044	2.5140*	2.77
5.	Knowledge about the department (X_{21})	0.0993	4.0471*	6.23

* = Significant at 1 per cent level of significance;
** = Significant at 5 per cent level of significance.
R^2 Value = 0.21; *F* Value = 8.89* with 5 and 174 d.f.

Table 8.58 : Results of stepwise regression analysis (backward elimination method) of independent variables of dairy farmers with their extent of overall functional linkage with extension personnel

Sl. No.	*Independent Variables*	*b-Values*	*t-Values*	*Relative Contribution (%)*
1.	Age (X_1)	-0.1059	2.6943*	9.56
2.	Family type (X_4)	-1.7756	1.9694**	5.00
3.	Social participation (X_{14})	1.0187	2.2967**	9.29
4.	Extension contact (X_{15})	0.3768	6.2107*	22.61
5.	Knowledge about the department (X_{21})	0.3543	17.6130*	30.44

* = Significant at 1 per cent level of significance;
** = Significant at 5 per cent level of significance.
R^2 Value = 0.77; *F* Value = 114.57* with 5 and 174 d.f.

From Table 8.59, it could be observed that the variables of research personnel, like their age, professional experience, value orientation and external environment were significantly associated with their ECL with dairy farmers. On multiple regression analysis, it was found that three variables namely, professional experience, value orientation and external environment affected researchers' ECL with dairy farmers significantly. Jointly, selected 14 independent variables made 57 per cent contribution to the variation of ECL (Table 8.59). On stepwise regression analysis, in final model, the above three variables retained which jointly made 47 per cent contribution to the ECL.

Table 8.59 : Zero order correlation coefficients and multiple regression coefficients of selected independent variables of research personnel with their extent of communication linkage with farmers

Sl. No.	*Independent Variables*	*b-Values*	*t-Values*	*Relative Contribution (%)*
1	2	3	4	5
1.	Cadre (X_2)	-0.3671**	-4.3941	0.5964
2.	Age (X_3)	0.2497	0.4747	0.2827
3.	Education (X_4)	0.2186	-0.4321	0.0495
4.	Professional experience (X_5)	0.2314	-1.5256	1.0635
5.	Training (X_6)	-0.0385	0.6291	0.7852
6.	Attitude (X_7)	0.0193	0.2352	0.2923
7.	Family background (X_8)	-0.0818	-0.1856	0.1910
8.	Achievement motivation (X_9)	0.0033	-0.0241	0.0184
9.	Value orientation (X_{10})	0.2519**	2.7370	2.4399
10.	Job satisfaction (X_{11})	-0.1439	-0.0804	0.1452
11.	Morale (X_{12})	0.1095	-0.0139	0.0250
12.	Perception of management (X_{13})	-0.0245	-0.3490	0.6006
13.	External variable (X_{14})	0.3123*	2.3095	1.7603
14.	Organizational climate (X_{15})	-0.0939	-0.3890	0.9934

* = At 1 per cent level of significance;
** = At 5 per cent level of significance.
R^2 Value = 0.57; *F* Value = 1.41* with 14 and 170 d.f.

Table 8.60 : Results of stepwise regression analysis (backward elimination method) of independent variables of research personnel with their extent of communication linkage with farmers.

Sl. No.	*Independent Variables*	*b-Values*	*t-Values*	*Relative Contribution (%)*
1.	Professional experience (X_5)	-1.1412	2.1553**	
2.	Value orientation (X_{10})	2.0761	2.5496**	
3.	External environment (X_{14})	2.5401	2.5881**	

* = At 1 per cent level of significance; ** = At 5 per cent level of significance.
R^2 Value = 0.47; *F* Value = 5.92* with 3 and 27 d.f.

Table 8.61 : Organizational constraints as perceived by the research and extension personnel in maintaining linkages between them.

Sl.No.	*Constraints*	*Research Personnel*		*Extension Personnel*	
		Pooled (n=32)	*Rank*	*Pooled (n=47)*	*Rank*
1.	Lack of provisions of joint activities/programmes for linkage	**24(75.00)**	II	**38(80.85)**	II
2.	Lack of formal mechanism for transfer of research findings to the extension agencies	**27(84.38)**	I	**42(89.36)**	I
3.	Organization set-up is rules oriented rather than programmes oriented	**19(59.37)**	V	**34(72.34)**	IV
4.	Lack of appropriate administrative policy of research and extension activities	**17(53.13)**	VI	**29(61.70)**	VI
5.	Lack of provision of compulsory involvement in joint activities	**21(65.62)**	IV	**26(55.32)**	VII
6.	Biasedness towards on-station researches	**13(40.63)**	VII	**36(76.59)**	III
7.	Too much bureaucracy and highly centralised administrative control	**23(71.87)**	III	**30(63.83)**	V

Figures in parentheses indicate percentage.

Since, dairy farmers' extent of linkage with researchers was found absent in the sample, hence, their selected variables were not subjected for this type of statistical analyses. The reasons for absence of linkage of dairy farmers with researchers, hence, was studied in terms of the constraints perceived by them. This aspect has been discussed in the subsequent subheads.

8.4 Constraints in Linkage between Research, Extension and Dairy Farmers

The third dimension of the present investigation was to delineate the bottlenecks as perceived by the selected respondents, constraining linkages among them. In the preceding heads, it was found that status of linkage differed between three continuum of interaction, *viz.*, research extension, extension farmers and research farmers. Further, except in case of farmer extension linkage, selected variables could predict little to linkage strength. Hence, attempt was made to identify and delineate the constraints which might be affecting the extent of functional linkage among the three selected partners of development.

8.4.1 Constraints in Linkage Between Research and Extension Personnel

Constraints as perceived by research and extension personnel in maintaining linkage between them were identified in the various areas by putting open ended questions to them. Based on the response, frequency distribution was done. Findings have been presented and discussed in the following subheads :

8.4.1.1 *Organizational Constraints*

Findings of Table 8.61 indicate that above 84 per cent of the RP and EP felt the lack of formal mechanism for transfer of research findings to the extension agencies/departments. Second important organizational constraint as perceived by both of them (above 75%) was the absence of provision of joint activities/ programmes for linkage. Similarly, about 77 per cent of EP viewed that there was biasedness towards on station researches and about 72 per cent of RP identified too much bureaucracy on research, thus affecting the linkage activities. The other organizational problems perceived by EP and RP were excessive rule orientedness in the organization, absence of appropriate administrative policy of research and extension activities and no provision of compulsory involvement in joint activities in the order of severity. Findings are in the line of those reported by Coulter (1983) and Bennett (1988).

8.4.1.2 *Communication Constraints*

Findings in Table 8.62 reveals that whereas majority of RP (81.25%) gave first rank to the absence of well defined channels of communication between research and extension, about 72 per cent of EP felt that even the available channels of communication performed far from satisfactory. They also felt that their representation in research and extension activities was limited and they lacked adequate knowledge about the pheno-mena of linkage.

Table 8.62 : Communication constraints as perceived by the research and extension personnel in maintaining linkages between them.

Sl.No.	*Constraints*	*Research Personnel*		*Extension Personnel*	
		Pooled (n=32)	*Rank*	*Pooled (n=47)*	*Rank*
1.	Absence of well defined channels of communication between research and extension	26 (81.25)	I	33 (70.21)	I
2.	Available communication channels are working far from satisfactory	19 (59.37)	III	34 (72.34)	I
3.	Poor knowledge about the research-extension linkage phenomena	17 (53.12)	IV	21 (44.68)	III
4.	Limited representation in research/ extension activities	22 (68.75)	II	34 (72.34)	I

Figures in parentheses indicate percentage.

8.4.1.3 *Budgetary/Financial Constraints*

The important budgetary problems felt by the sampled personnel were inadequacy of budget for carrying out joint activities, injudicious allocation of funds for research and extension activities, late reimbursement of even externally aided projects and lack of financial encouragement to the field oriented researches in that order (Table 8.63).

Table 8.63 : Budgetary/financial constraints as perceived by the research and extension personnel in maintaining linkages between them.

Sl.No.	*Constraints*	*Research Personnel*		*Extension Personnel*	
		Pooled (n=32)	*Rank*	*Pooled (n=47)*	*Rank*
1.	No provision of adequate budget for carrying joint activities	28 (87.50)	I	36 (76.59)	I
2.	Injudicious allocation of funds for the research and extension activities	17 (53.12)	III	28 (59.57)	II
3.	Externally aided funds are even not reimbursed timely	22 (68.75)	II	15 (31.91)	III
4.	Field oriented researches are not encouraged financially	16 (50.00)	IV	13 (27.66)	IV

Figures in parentheses indicate percentage

8.4.1.4 *Psychological/Motivation Constraints*

Above 72 per cent of both RP and EP felt that there was no performance oriented rewards. They also felt that the motivating force to pursue linkage related activities was absent in the organization (Table 8.64). Majority of them (above 53%) also opined that there was lack of interest for coordinated research and extension activities. Moreover, lack of due appreciation and recognition for any good work and clash of personality and cadre among them were yet another important psychological constraints in the interaction of research and extension personnel.

8.4.1.5 *Technological Constraints*

Findings in Table 8.65 reveals that as high as 81 per cent of EP doubted the appropriateness of the available transferable technologies owing to the basic flaws in planning and implementation of the research agenda. The same point was also highlighted by majority of EP, but with lesser frequency (65.62%). Perhaps, this was the reason that about 74 per cent of EP said the limited availability of transferable dairy production technologies. Both RP and EP agreed on the point that research activities yielded only the findings to be published not the technology in concrete form to solve field level problems. Almost equal per cent (about 62%) of researchers and EP opined that there was no matching between farm level problems and recommended technological options.

Table 8.64 : Psychological/motivational constraints as perceived by the research and extension personnel in maintaining linkages between them.

Sl.No.	*Constraints*	*Research Personnel*		*Extension Personnel*	
		Pooled (n=32)	*Rank*	*Pooled (n=47)*	*Rank*
1.	Absence of motivating force to think about linkage related activities	22 (68.75)	II	34 (72.34)	II
2.	Absence of performance oriented rewards	23 (71.87)	I	35 (74.46)	I
3.	Lack of appreciation and recognition for any good work	17 (53.12)	IV	21 (44.68)	V
4.	Lack of interest for coordinated research and extension activities	19 (59.37)	III	25 (53.19)	III
5.	Personality and cadre clash	13 (40.62)	V	22 (46.81)	IV

Figures in parentheses indicate percentage

From the findings presented in the above subheads, it could be said that the poor status of linkage strength between research and extension could be attributed to the most important bottlenecks such as lack of formal linkage mechanism as well as the joint activities for linkage. This clearly indicates the existing structural linkage mechanisms need to be overhauled. The frequency of activities of these mechanisms need to be increased with greater provision of participation of research and extension personnel of various cadre. The above types and frequency of constraints were also reported by Compton (1984), Balaguru and Rajgopalan (1986) and Singh (1993).

Table 8.65 : Technological constraints as perceived by the sampled research and extension personnel in maintaining linkages between them.

Sl.No.	*Constraints*	*Research Personnel*		*Extension Personnel*	
		Pooled (n=32)	*Rank*	*Pooled (n=47)*	*Rank*
1.	Limited availability of transferable appropriate technologies	15 (46.87)	IV	35 (74.47)	II
2.	Available technologies are not appropriate for field level application owing to the basic flaws in planning and implementation of research agenda	21 (65.62)	I	38 (80.85)	I
3.	Research activities output only the findings to be published not the technology in the concrete form	19 (59.37)	III	32 (68.08)	III
4.	No matching of farmers problems and technological options recommended	20 (62.50)	II	29 (61.70)	IV

Figures in parentheses indicate percentage

8.4.2 Constraints in Linkage Between Extension Personnel and Dairy Farmers

The linkage constraints between extension personnel and dairy farmers were ascertained as perceived by both of them.

8.4.2.1 *Linkage Constraints as Perceived by the Extension Personnel*

Under the following subheads, these constraints are delineated and discussed :

8.4.2.1.1 Organizational Constraints

Findings in Table 8.66 indicates that majority of EP (79%) felt that there was weak institutionalized mechanism to involve

Table 8.66 : Organizational constraints perceived by the extension personnel in maintaining linkages with the dairy farmers.

Sl.No.	*Constraints*	*Extension Personnel*				
		NDRI (n=16)	*HAU (n=15)*	*SDAH (n=16)*	*Polled (n=47)*	*Rank*
1.	Lack of adequate mechanism to promote interaction between extension and farmers	12 (75.00)	9 (60.00)	8 (50.00)	29 (61.70)	IV
2.	Absence of the educational concept in animal husbandry and dairy extension	10 (62.50)	8 (53.33)	13 (81.25)	31 (65.96)	II
3.	The existing extension strategy is unable to meet farmers expectations/problems	8 (50.00)	9 (60.00)	10 (62.50)	27 (57.45)	V
4.	Weak institutionalized mechanism to involve farmers in planning and decision making process	12 (75.00)	11 (73.33)	14 (87.50)	37 (78.72)	I
5.	Problems of coordination between the field level activities and headquarters	9 (56.25)	10 (66.67)	9 (56.25)	28 (59.57)	VI
6.	Lack of farming systems approach	8 (50.00)	10 (66.67)	12 (75.00)	30 (63.83)	III
7.	Too much administrative interference and heavy paper works	10 (62.50)	9 (60.00)	11 (68.75)	30 (63.83)	III
8.	Delayed decision making due to centralised power	8 (50.00)	8 (53.33)	7 (43.75)	23 (48.94)	VII

Figures in parentheses indicate percentage

Table 8.67 : Budgetary constraints perceived by the extension personnel in maintaining linkages with the dairy farmers.

Sl.No.	*Constraints*	*Extension Personnel*				
		NDRI (n=16)	*HAU (n=15)*	*SDAH (n=16)*	*Polled (n=47)*	*Rank*
1.	Inadequacy of fund/budget for field activities	13 (81.25)	13 (86.67)	14 (87.50)	40 (85.11)	I
2.	Delayed reimbursement of contingency, travelling and daily allowance	6 (37.50)	10 (66.67)	12 (75.00)	28 (59.57)	II
3.	Diversion of funds for other administrative works	8 (50.00)	9 (60.00)	10 (62.50)	27 (57.45)	III

Figures in parentheses indicate percentage

Table 8.68 : Mobility/conveyance constraints perceived by the extension personnel in maintaining linkages with the dairy farmers.

Sl.No.	*Constraints*	*Extension Personnel*				
		NDRI (n=16)	*HAU (n=15)*	*SDAH (n=16)*	*Polled (n=47)*	*Rank*
1.	Non-availability of vehicle as and when required for the field visits/activities	6 (37.50)	10 (66.67)	12 (75.00)	28 (59.57)	II
2.	Too much official procedure in getting the vehicle sanctioned for field activities	10 (62.50)	12 (80.00)	8 (50.00)	30 (63.39)	I
3.	Field level workers have to rely on their own vehicle for visiting the areas of operations	8 (50.00)	6 (40.00)	10 (62.50)	24 (51.06)	III

Figures in parentheses indicate percentage

farmers in planning and decision making process regarding extension activities. About 66 per cent of them also agreed on the fact that there was absence of educational concept in animal husbandry and dairy extension. Other organizational constraints as perceived by not less than 49 per cent of EP were lack of farming systems approach, lack of adequate mechanism to promote interaction between extension and farmers, inability of existing extension strategy to meet farmers' problems and poor coordination between field level activities and headquarters in that order.

8.4.2.1.2 Budgetary Constraints

Three important budgetary constraints as perceived by not less than 57 per cent of EP were inadequacy of budget/funds for field activities, delayed reimbursement of contingency, travelling and daily allowances and diversion of funds for other administrative works in that order (Table 8.67).

8.4.2.1.3 Mobility/Conveyance Constraints

Mobility problem was relatively less frequently felt by the EP in order to link them with dairy farmers. Still, 63 per cent of them opined that there was much official procedure in getting the vehicle for field activities. Similarly, 60 per cent of EP stated the non availability of vehicle as and when required for the field visit. Due to above two reasons, perhaps, 51 per cent of sampled EP said that the field level workers had to depend most on their own vehicle for visiting the areas of operation (Table 8.68).

8.4.2.1.4 Motivational Constraints

Some of the constraints which could not sufficiently motivate EP to link with dairy farmers were identified and are presented in Table 8.69. Findings of the table reveal that the lack of due incentives for the field extension work and no suitable policy for transfer, reward and punishment were first two bottlenecks, respectively, which demotivated the EP. Moreover, more than half of the EP were found bored with routined field activities and hence, they lacked sufficient enthusiasm. This fact is further comprehensible from the another problem, *i.e.*, field works achievement was little to do with career development as felt by about 43 per cent of them.

8.4.2.1.5 Technical Constraints

A perusal of Table 8.70 indicates that most frequently felt technical constraint was the lack of handy diagnostic kits to be used at field level (68%). Similarly about 60 per cent of EP found that there was lack of stationery for publication of leaflets, handouts, newsletter, etc. in order to make and maintain linkages with the farmers. Two more problems, *viz.*, lack of more appropriate technologies as per farmers' problems and inadequacy of technical inputs like semen, vaccines, medicines, etc. were accorded third rank by 57 per cent of the EP. More than half of them, particularly from HAU extension system, felt the need of more numbers of field level veterinary institutions.

8.4.2.1.6 Farmers' Related Constraints

Findings in Table 8.71 reveals that 74 per cent of the EP felt that farmers were more interested in free supply of inputs from the department and lesser willing to attend meeting/ demonstration, trials, training, etc. Farmers' preference to crop related activities was yet another linkage constraints in dairying. Similarly, about 62 per cent of EP were of the view that groupism among farmers and wide heterogeneity among them constrained the EP to devise any uniform development plan. Moreover, farmers were also little willing to reveal their resources, practices and problems, hence the effective linkage between them suffered.

Findings lead to inference that the linkage of EP with dairy farmers can not be strengthened in the want of appropriate organizational mechanism to bring both together. A number of budgetary and psychological bottlenecks constrained the linkage strength. Findings are in the line of those reported by Malik (1993) and Bharati (1993).

8.4.2.1.6 Farmers' Related Constraints

Findings in Table 8.71 reveals that 74 per cent of the EP felt that farmers were more interested in free supply of inputs from the department and lesser willing to attend meeting/ demonstration, trials, training, etc. Farmers' preference to crop related activities was yet another linkage constraints in dairying. Similarly, about 62 per cent of EP were of the view that groupism among farmers and wide heterogeneity among them constrained the EP to devise any uniform development plan. Moreover, farmers were also little willing to reveal their resources, practices and problems, hence the effective linkage between them suffered.

Findings lead to inference that the linkage of EP with dairy farmers can not be strengthened in the want of appropriate organizational mechanism to bring both together. A number of budgetary and psychological bottlenecks constrained the linkage strength. Findings are in the line of those reported by Malik (1993) and Bharati (1993).

8.4.3 Constraints in Linkage Between Research Personnel and Dairy Farmers

The problems in maintaining reciprocal interaction between research and farmers were ascertained as viewed by research personnel and dairy farmers. Under the following sub heads results have been presented and discussed:

8.4.3.1 *Linkage Constraints as Perceived by Research Personnel*

Discussion with Research Personnel (RP) reveals that majority of the RP (84.44%) were pre-occupied with on station works and hence, they had little time for field activities. More than two third of them also felt that there was poor provision and mechanism for on farm farmers participatory research. Owing to it, above 70 per cent of RP did not have technological feedback from farmers. Moreover, about 69 per cent of RP opined that research manager of the organization never evaluated the progress of RP on the basis of their involvement in field activities. Few more problems as perceived by the RP were poor resource status of dairy farmers, non availability of farmers at the time of visit, language problem and prevalence of groupism and factionalism among villages.

Table 8.69 : Motivational/psychological constraints perceived by the extension personnel in maintaining linkages with the dairy farmers.

Sl.No.	*Constraints*	*Extension Personnel*				
		NDRI (n=16)	*HAU (n=15)*	*SDAH (n=16)*	*Polled (n=47)*	*Rank*
1.	Lack of due incentives for the field extension works	10 (62.50)	12 (80.00)	11 (68.75)	32 **(68.08)**	I
2.	Lack of suitable policy for transfer, reward and punishment	9 (56.25)	9 (60.00)	10 (62.50)	28 **(59.57)**	II
3.	Routined field works are often unable to generate any enthusiasm	8 (50.00)	7 (46.67)	9 (56.25)	24 **(51.06)**	III
4.	Field works achievement has little to do with promotion	7 (43.75)	7 (46.67)	6 (37.50)	20 **(42.55)**	IV
5.	Inadequate facilities discourages the field level works	4 (25.00)	7 (46.67)	5 (31.25)	16 **(34.04)**	V

Figures in parentheses indicate percentage

Table 8.70 : Technical constraints perceived by the extension personnel in maintaining linkages with the dairy farmers.

Sl.No.	*Constraints*	*Extension Personnel*				
		NDRI (n=16)	*HAU (n=15)*	*SDAH (n=16)*	*Polled (n=47)*	*Rank*
1.	Lack of more appropriate technologies which could match farmers' problems	9 (56.25)	8 (53.33)	10 (62.50)	27 (57.44)	III
2.	Lack of handy diagnostic kits to be used at field level	10 (62.50)	12 (80.00)	10 (62.50)	32 (68.08)	I I
3.	Inadequacy of technical inputs, *viz.*, semen, vaccines, medicines, etc.	8 (50.00)	10 (66.67)	9 (56.25)	27 (57.45)	III
4.	Lack of stationery for publication of leaflets, handouts, newsletters, etc.	7 (43.75)	10 (66.67)	11 (68.75)	28 (59.57)	II
5.	Lack of more numbers of field level veterinary institutions	6 (37.50)	12 (80.00)	6 (37.50)	24 (51.06)	IV
6.	Lack of audio-visual aids for making effective communication with the farmers	5 (31.25)	6 (40.00)	6 (37.50)	17 (36.17)	V

Figures in parentheses indicate percentage

8.4.3.2 *Linkage Constraints as Perceived by Dairy Farmers*

Table 8.73 reflects that as high as about 87 per cent of the dairy farmers denied the visit of researchers to the villages. Secondly, they were also not aware of their role in dairying research and development. Half of the DFs said that they were unable to distinguish between EP and RP if any joint visit is made by them. A considerable percentage (39%) of DF blamed that escapism prevailed in the organization if the Institute is visited by them. The findings are in the line of the same reported by Eponou (1996), Fischer *et al.* (1996) and Farrington (1997).

Based on the findings described and discussed in this chapter, it could be comfortably concluded that the strength of functional linkage among research, extension and dairy farmers showed varied status on the possible three continuum. Whereas, linkage between dairy research personnel and extension personnel was absent to weak, linkage between extension personnel and dairy farmers displayed more or less satisfactory picture. The antecedent variables could speak little in predicting the status of linkage among them except in case of dairy farmers. Further, it could be derived that the constraining item like absence of the in built mechanism with the organization/department to combine and promote reciprocal interaction among them was the major limiting factor to the linkage strength. The empirical model of present investigation of research-extension-farmers linkage has been depicted as in Figure 2.

Table 8.71 : Farmers related constraints perceived by the extension personnel in maintaining linkages with the dairy farmers.

Sl.No.	Constraints	Extension Personnel				
		NDRI (n=16)	HAU (n=15)	SDAH (n=16)	Polled (n=47)	Rank
1.	Farmers are more interested in free supply of inputs and lesser willing to attend to meeting/demonstration, trials, training, etc.	12 (75.00)	12 (80.00)	11 (68.75)	35 **(74.46)**	I
2.	More preference to crop related activities	10 (62.50)	11 (73.33)	9 (58.25)	30 **(63.83)**	II
3.	Non cooperation during field extension activities	8 (50.00)	9 (60.00)	10 (62.50)	27 **(36.17)**	VII
4.	Groupism/factionalism among farmers which sometimes makes difficult to operate any group activities	9 (56.25)	8 (53.33)	9 (58.25)	26 **(55.32)**	IV
5.	Wide heterogeneity among farmers create problems in making any uniform development plan	10 (62.50)	9 (60.00)	10 (62.50)	29 **(61.70)**	III
6.	Farmers shows inhibitions in showing/revealing their resources, practices, problems, etc.	6 (37.50)	8 (53.33)	7 (43.75)	21 **(44.68)**	V
7.	Farmers are not willing to understand the technicalities involved in new information	7 (43.75)	6 (40.00)	6 (37.50)	19 **(40.43)**	VI

Figures in parentheses indicate percentage

Table 8.72 : Constraints perceived by the dairy farmers in maintaining linkages with the extension personnel.

Sl.No.	*Constraints*	*Extension Personnel*				
		NDRI (n=60)	*HAU (n=60)*	*SDAH (n=60)*	*Polled (n=180)*	*Rank*
1.	Department operates through few selected and rich farmers of the village	35 (58.33)	40 (66.67)	32 (53.33)	107 (59.44)	VII
2.	Farmers are rarely invited for planning and decision making about field extension activities	40 (66.67)	50 (83.33)	45 (75.00)	135 (75.00)	II
3.	There is little matching between farm level problems and technologies offered by the department	42 (70.00)	52 (86.67)	40 (66.67)	134 (74.44)	III
4.	Lesser frequency of meeting, camp, campaign, calf rallies, etc.	45 (75.00)	55 (91.67)	50 (83.33)	150 (83.33)	I
5.	Information offered are more of theoretical nature	40 (66.67)	48 (80.00)	42 (70.00)	130 (72.22)	IV
6.	Non-availability of the experts at the time of visit to the centre	25 (41.67)	50 (83.33)	30 (50.00)	105 (58.33)	VIII
7.	Personnel are less interested in understanding the entire farm problems	20 (33.33)	35 (58.33)	36 (60.00)	91 (50.56)	IX
8.	Department helps only in the form of inputs and services and not in the educational form	30 (50.00)	10 (16.67)	50 (83.33)	90 (50.00)	X
9.	Little provision of field level training to the farmers	35 (58.33)	25 (41.67)	30 (50.00)	90 (50.00)	X

(Contd...)

Sl.No.	Constraints	Extension Personnel				
		NDRI (n=60)	*HAU (n=60)*	*SDAH (n=60)*	*Polled (n=180)*	*Rank*
10	Training needs are not assessed from time to time	30 (50.00)	40 (66.67)	45 (75.00)	115 (63.89)	VI
11.	Training imparted at the centre are more of theoretical nature	25 (41.67)	10 (16.67)	20 (33.33)	55 (30.56)	XI
12.	The inputs are not adequately available and heavy charges are levied for the services	35 (58.33)	50 (83.33)	40 (66.67)	125 (69.44)	V

Figures in parentheses indicate percentage

Table 8.73 : Constraints faced by the dairy farmers in maintaining linkages with the research personnel

Sl.No.	Constraints	Extension Personnel				Rank
		NDRI (n=60)	HAU (n=60)	SDAH (n=60)	Polled (n=180)	
1.	Not aware about researchers' role in dairy development	43 (71.67)	48 (80.00)	50 (83.33)	141 **(78.33)**	II
2.	No visit of researchers to the villages	50 (83.33)	52 (86.67)	54 (90.00)	156 **(86.67)**	I
3.	Unable to distinguish between extension personnel and research personnel during their joint visit/activity, if any	35 (58.33)	30 (50.00)	15 (25.00)	90 **(50.00)**	III
4.	Escapism prevails in case if the centre/ institute is visited	20 (33.33)	125 (58.42)	10 (16.67)	55 **(38.55)**	IV
5.	Research recommendations are not applicable to the existing farming conditions	18 (30.00)	22 (36.67)	12 (20.00)	52 **(28.89)**	V

Figures in parentheses indicate percentage

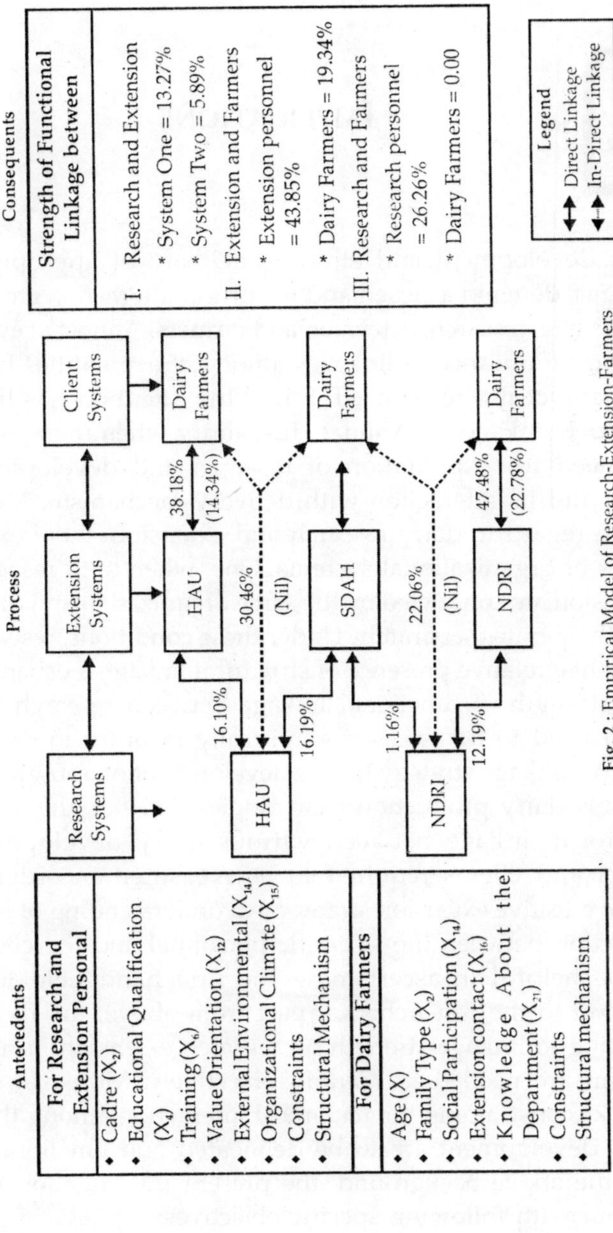

Fig. 2 : Empirical Model of Research-Extension-Farmers Linkage (figures in parentheses indicate farmers' status of linkage)

9 IMPLICATIONS

The development and effective delivery of appropriate technologies demand a closed and reciprocal interaction among the entities like, research, extension and farmers. Almost at every forum, the significance of linkages among above entities have been unequivocally recognized. Still, little attention has been paid on such linkages in Animal Husbandry when there exists a great need for exploitation of research and development activities, and its interaction with delivery mechanism. Moreover, with regard to dairy research and extension, there exists two types of organizational systems. One, when both research and extension are conducted by the same organization and other, when both operates separately. Under these conditions, research questions like relative presence of structural linkage mechanism and the strength of functional linkage between research and extension need to be assessed empirically in order to design the effective linkage strategy for the development of appropriate transferable dairy production technologies. Further, the status of reciprocal linkage between various dairy development personnel and clients required to be measured in order to formulate effective extension strategy. An understanding of level of interaction between the scientific personnel and the clients would be helpful in ascertaining the first hand field level information to the researchers. Apart from above, the factors contributing to linkage strength on research extension farmers continuum are needed to identify objectively. Moreover, the bottlenecks constraining the reciprocal interaction among these actors of development are to be delineated and ameliorated. Against the above background, the present investigation was carried out with following specific objectives:

(*i*) To develop the indices and measure the extent of structural and functional linkages among research, extension and clients.

(*ii*) To identify the factors influencing the strength of functional linkage on research farmers continuum.

(*iii*) To delineate the linkage constraints perceived by researchers, extension personnel and farmers.

Study was conducted in the purposively selected state of Haryana. Three organizations, *viz.*, NDRI, Karnal; HAU, Hisar and SDAH, Karnal were included in the study. Three categories of respondents as research personnel (32), extension personnel (47) and dairy farmers (180) were taken in order to meet the requirements of objectives. Dairy farmers were sampled by stratified proportionate random sampling technique from the nine adopted villages under the three selected extension systems from the purposively selected Karnal district. The research variables were measured by using suitable scales and by devising the schedules and indices. The collected data were coded, tabulated and subjected to the applicable statistics.

9.1 Salient Findings

1. Very few structural linkage mechanism (SLM) was found developed by both NDRI and HAU for Research Extension (RE) linkage. The frequency of activities of these mechanisms was quite less. However, HAU outperformed NDRI in this respect both with their personnel as well as with SDAH personnel. The extension personnel from SDAH reciprocated better with HAU than with NDRI on structural arrangements. The three selected extension systems were found to have good number of SLM for the dairy farmers with considerable variation in their frequency and status of activities.

2. The Research Extension Functional Linkage (REFL) in both the selected organizational system was found to be absent to weak on all the parameters, *viz.*,

communication, collaborative professional activities, planning and decision making, implementation and evaluation, training and supply and services. However, at the existing strength of REFL, the organizational system containing both research and extension together performed better (13.27%) than when both were separated (5.89%). Further, HAU significantly outperformed NDRI with respect to the strength of REFL both within the organization and with SDAH.

3. The strength of functional linkage between extension personnel (EP) and dairy farmers was stated as moderate (28.75, 60.45) by most of the EP (72.34%). On this count, NDRI was better than HAU and SDAH. However, HAU had edge over NDRI and SDAH in linking with dairy farmers with respect to communication linkage and joint planning and deciding the field programmes. Still, except for supply and services, no significant difference was noted in the strength of other parameters of functional linkage of EP of selected extension systems with dairy farmers. Whereas, SDAH and NDRI extension systems were better linked with dairy farmers with respect to supply and services, HAU did well in terms of planning and decision making and training.

4. From dairy farmers' points of view, their strength of functional linkage with EP was to the extent of 19.34 per cent and a considerable gap of about 52 per cent was noted in the linkage between them. This gap was significantly higher (between 49.51 to 60.32%) on all the parameters except in supply and services (25.22%). The variation of the strength of functional linkage between selected categories of dairy farmers was found non significant in all the selected parameters except in communication, implementation and evaluation.

5. Researchers' extent of communication linkage with dairy farmers was computed to the extent of 30.26 per cent. With this regard, no significant difference was found

between NDRI (22.06%) and HAU (30.46%). However, sampled dairy farmers stated no interaction with researchers.

6. The strength of research extension functional linkage (REFL) was found significantly affected by the variables, like, cadre, educational qualification, training received and attitude of the personnel. Fourteen selected independent variables, however, could explain 50 per cent in the variation of REFL. On stepwise regression analysis, except attitude, above three variables retained in the final model, which jointly predicted 35 per cent in the variation of REFL.
7. The variables of EP, *viz.*, training, value orientation, job satisfaction and organizational climate had significant bearing on their strength of functional linkage with the clients. Jointly, all the selected variables (15) could explain 50 per cent in the variation of functional linkage. On stepwise regression analysis, except job satisfaction, remaining above three variables were responsible for 30 per cent variation.
8. The strength of linkage of dairy farmers with the EP were significantly influenced by the factors, like, age, family type, social participation, extension contact and knowledge about the department. Altogether, 21 independent variables of farmers were responsible for 78 per cent in the variation of the linkage strength. In the final model, the above five variables predicted to the extent of 77 per cent.
9. The strength of communication linkage of researchers with dairy farmers was found to be significantly influenced by the three variables, like, professional experience, value orientation and external environment. These three variables in final model explained 47 per cent variation in the linkage strength.
10. Most important research extension linkage constraint identified was the absence of formal mechanism for constant and regular interaction between research and

extension personnel. Further, a number of organizational, budgetary, motivational and technical bottlenecks were identified which affected the RE linkage strength.

11. Weak institutionalized mechanism to involve farmers in planning and designing of extension activities was perceived constraints by the majority of EP. The extension personnel linkage with farmers was constrained by number of organizational, budgetary, mobility, motivational and technical bottlenecks. On the other hand, farmers felt that there were lesser frequency of meeting, camp, campaign, etc. and they were rarely invited for planning and decision-making by the departments about field extension activities of farmers' interest and needs.

12. Similarly, researchers pre-occupation with on-station research works and poor provision and mechanism for on farm farmers participatory research were identified as the most frequently perceived constraints affecting the linkage of researchers with the farmers. Farmers, on the other hand, denied the visit of researchers to the villages. At the same time, farmers were found not aware of their role in dairying research and development.

9.2 Major Recommendations

Based on the findings of the present investigation, following implications could be recommended for research and extension manager as well as the policy makers in general and for those of the selected organizations/departments in particular:

1. Study reveals that there is greater need for institutionalization of more number of structural linkage mechanism especially the technology treatment group/committee for better linkage between Research and Extension (RE) personnel in dairying.

2. Frequency of activities of existing structural mechanisms needs to be increased with greater participation of RE personnel of different cadre.

3. As the functional linkage between RP and EP was observed absent to weak on all the selected parameters, following three tier functional linkage mechanism could be intervened for RE personnel to facilitate joint planning and decision making, implementation and evaluation of collaborative RE activities :

Sl. No.	*Level of RE Personnel*	*Activities*
1.	Top Level	Designing, monitoring and final evaluation of research and extension activities with the participation of middle and lower level RE personnel.
2.	Middle Level	Implementation and monitoring of research/extension activities with lower and top level RE personnel.
3.	Lower Level	Collaborative and supportive role with middle and top level RE personnel

4. A periodical and regular training programme for extension personnel ought to be conducted and strengthened by the research organizations/department in order to transfer appropriate information to the EP for the effective dissemination of same to the clients.

5. Extension farmers functional linkage requires to be strengthened on the parameters, *viz.*, communication, joint planning and decision making regarding field extension activities and training. The present status of linkage of extension personnel in supply and services with the farmers need to be maintained. The selected extension systems must make a balanced approach of both educational as well as supply and services for effective extension farmers linkage.

6. As the strength of research extension linkage was found to be affected by the variables like cadre, educational qualification and attitude of the personnel, the research and extension managers of the selected organizations/departments must consider the importance of these aspects.

7. The factors identified, *viz.*, training, job satisfaction and organizational climate of the extension department as

well as EP need to be properly manipulated by the extension managers to improve the extension-farmers' linkage.

8. The selected extension departments require to be further intensive in their functions so that dairy farmers may have better extension contact and knowledge about the department. This, in turn, will improve the strength of farmers' linkage with extension personnel.
9. The constraints identified needs to be debottlenecked by the top level research and extension managers as well as by the policy makers in order to improve the linkages among farmers, extension personnel and researchers.
10. More number of on farm farmers participatory researches are required to be designed in order to bring farmers, extension personnel and researchers at common platform.
11. For different types of technologies/practices, different linkage mechanisms need to be devised.

9.3 Future Research Suggestions

As no empirical study could be made perfect closed ended, the present investigation was also not the exception. The experiences of the study explored some of the opportunities which could be investigated in future :

1. There is greater scope of making comparative study of such linkages in animal husbandry *vis a vis* agriculture.
2. An extensive study on linkages could be formulated by including more number of actors of development, *viz.*, input supplier, financial institutions, village organizations, etc. and the flow of knowledge and information could be mapped.
3. Such type of study could also be contemplated by adopting qualitative approach as well as case study technique.

LITERATURE CITED

Abedin, M.Z. and Chowdury, M.K. (1989). Organizational and managerial innovations for linking extension and research in Bangladesh. Paper prepared for the International Workshop on Making the Link between Agricultural Research and Technology Users, ISNAR, The Hague, 19–25 November.

Acharya, R.M. (1994). Welcome and introductory remarks in the national seminar on Extension Education and Research and Development Linkage, held at Conference Hall, IARI Library, New Delhi.

Akhouri, M.M.P. (1973). Communication behaviour of extension personnel: An analysis of Haryana agricultural extension system. Ph.D. Thesis, I.A.R.I., New Delhi.

Alzahrwal, K.H. (1992). Utilization of extension methods and aids and their importance in transferring agricultural knowledge and skills as viewed by extension workers: A field study in the central region of Saudi Arabia. Department of Agriculture, King Saud University, Saudi Arabia, World Agril. Econ. & Rural Sociology Abstr., *34*(10): 866.

Ambastha, C.K. (1974). Communication patterns in farm innovation development. Extension and Client Systems in Bihar : A Systems Approach. Ph.D. Thesis, I.A.R.I., New Delhi.

Ambastha, C.K. (1980). Communication pattern of farm scientists. *Indian J. Extn. Edn.*, *16*(1&2): 34–38.

Ambastha, C.K. (1986). Communication pattern in innovation development, extension and client systems (a system approach). Delhi: BR Publishing Corporation.

Ambastha, C.K. and Singh, K.N. (1976). Farm scientists communicate to farmers. *Indian J. Extn. Edn.*, *13*(3&4) : 1–5.

Ambastha, C.K. and Singh, K.N. (1977). Communication patterns of farm scientists : A system analysis. *Indian J. Extn. Edn., 13*(1&2) : 9–16.

Ambastha, C.K. and Singh, K.N. (1979). An analysis of inter system communication pattern. *Indian J. Extn. Edn., 15*(1&2): 1–8.

Anonymous (1969). Recommendations of the All India Conference on Agricultural Education held at Bangalore.

Anonymous (1970). A method of assessing progress of agricultural university. Joint Indo American Study Team Report, Part I, ICAR, New Delhi.

Anonymous (1992). Research-extension linkages. Decentralized effort. *Agril. Extn. Rev., 4*(3) : 10–13.

Anonymous (1993). Report of the National Seminar on NATP, New Delhi.

Anonymous (1997). Dairy India, Rekha Printers Pvt. Ltd., New Delhi.

Antholt, C.H. (1990). Agricultural extension. Some concepts to be used and some to be set aside: Issues to be discussed and ideas to be pursued. *J. Extn. Systems, 6*: 1.

Ashby, J. (1990). Small farmers' participation in the design of technologies. In: Altieri, M.A. and Hecht, S.B. (eds.), Agro ecology and Small Farm Development. Boca Raton, Florida: CRC Press.

Axinn, G.H. and Thorat, S. (1972). Modernising world agriculture : A comparative study of agricultural extension education systems. New York, Praeger Publishers, New Delhi, Oxford and IBH.

Axinn, G.H. (1988). Guide on alternative extension approaches. Agriculture Education and Extension Services, FAO, Rome.

Axinn, G.H. (1991). Potential contribution of FSD to institution development. Paper presented at Asian *Farming Systems Research in Extension Symposium,* 19–22 November.

Axinn, T. (1972). Comparative extension for modernising world agriculture. Przegar Publication.

Azad, R.N. (1975). Rural extension in India - Change agents past and present. *Kurukshetra, 24* : 13–17.

Babu, A.R. and Sinha, B.P. (1985). Communication behaviour of extension personnel with regard to modern rice technology. *Indian J. Extn. Edn., 21*(3&4) : 11–15.

Balaguru, T. and Rajagopalan, M. (1986). Management of agricultural research projects in India. Part 2. Research Productivity Reporting and Communication. *Agricultural Administration, 23* :

Balakrishna, B. (1997). Evaluation of dairy production practices in selected farming systems of Karnataka state. Ph.D. Thesis, NDRI Deemed University, Karnal.

Baweja, G.S. (1974). Training of extension personnel and farmers. In: Technology Transfer Systems and Constraints. Edited by D. Singh and J.S. Pandey. Nirbal Ke Balram Press, Kanpur, p. 57.

Bennett, C. (1988). Improving coordination of extension and research through the use of interdependency model. Bologna: 7th International Congress of Rural Sociology, Theme Session of Extension.

Bhagat, R. and Mathur, P.N. (1985). Mass media and changing life style of farm women in Delhi territory. *Indian J. Extn. Edn., 21*(3) : 37–41.

Bhanja, S.K. (1981). A study of socio economic and psychological correlates in organization and functioning of dairy cooperatives. Ph.D. Thesis, NDRI, Karnal.

Bharati, V. (1993). Dairy development in Haryana: A system analysis. Unpublished Ph.D. Thesis, CCS HAU, Hisar.

Bhattacharya, A.K. and Talukdar, R.K. (1996). Motivational climate in the Gram Sevak Training Centres. *Rural India*, July 1996, pp. 20–25.

Biggs, S.D. (1989). Resource poor farmers' participation in research: A synthesis of experiences from nine national agricultural research systems. *OFCOR Comparative Study Paper No.3, The Hague : ISNAR.*

Block, K. and Seegers, S. (1988). The research-extension linkage in the Southern region of Sri Lanka - An agricultural information system perspective. Washington: AUW, Unpublished M.Sc. Thesis.

Bourgeosis, R. (1989). Promoting integration through structural change: Making the link between agricultural research and technology users. *ISNAR Working Papers.*

Brayfield, A.H. and Rothe, H.F. (1951). An index of job satisfaction. J. Appl. Psychol., *35*(5) : 307 311.

Bunting, A.H. (1983). Fifty years of experimental agriculture. *Experimental Agriculture, 19* : 1–13.

Burmeister, L. (1985). State, society and agricultural research policy: The case of South Korea. Ph.D. Dissertation, Cornell University.

Byrnes, F.C. (1968). Some missing variables in diffusion research and innovations. Philippines *Sociological Review, 14*(4) : 242.

Cernea, M.M. (1981). Sociological dimensions of extension organization : The introduction of the T&V System in India. World Bank Reprint Series: No.196. International Experience for Strategies for Planned Change. Edited by Bruce R. Crouch and Shankariah Chamala, pp. 233–237.

Cernea, M.M.; Coulter, J.K. and Russell, J.F.A. (1985). Research-extension farmer: A two way continuum to agricultural development. Washington, DC: The World Bank.

Chennegowda, M.B. (1983). Interaction between researchers and farmers in Bangalore district of Karnataka. *Indian J. Extn. Edn., 19*(3&4) : 66–68.

Cho, Chae Yun (1996). Transformation of Korean farming: A success story of effective linkage. Asia Pacific Association of Agricultural Research Institutions, FAO Regional Office for Asia and the Pacific, Bangkok, pp. 12–29.

Chopra, V.L. (1992). Address to the National Seminar on R&D Linkage and Feedback in Agriculture Development at Krishi Bhavan, New Delhi.

Claar, J.B. and Bentz, R.P. (1984). Organizational design and extension administration. In : B.E. Swanson (ed.). Agricultural Extension : A Reference Manual, Rome: F.A.O. of the United Nations.

Clausen, A.W. (1984). Towards sustainable development in sub Saharan Africa : A joint programme of action. Washington, DC : The World Bank.

Compton, J.L. (1984). Linking scientist and farmer. Rethinking extension's role. World Food Issues (2nd edition), Ithaca, N.Y.: Cornell University Centre for the Analysis of World Food Issues. Programme in International Agriculture.

Coughenour, M.C. (1968). Some general problems in diffusion from the perspective of the theory of social action. Diffusion Research Needs, North Central Regional Research Bulletin, *186*, University of Missouri, USA.

Coulter, J.K. (1983). The interdependence of research and extension. Agricultural Extension by Training and Visit. The Asian Experience, Washington, D.C. : World Bank.

Cummings, R.W. Jr. (1981). Strengthening linkages between agricultural research and farmers - An overview. Paper presented as a background document for Workshop on Linkages between Agricultural Research and Farmers in Developing Countries. New York: Rock Feller Foundations.

Das, P. (1996). Changing scenario of agricultural economy and future thrust in extension. *J. Extn. Edn., 7* : 1390–1397.

deJanvry, A. and LeVeen, E.P. (1983). Aspects of the political economy of technical change in developed economies. In: Technical Change in Social Conflict in Agriculture: Latin American Perspectives (M. Pineiro and E. Trigo, eds.), Westview Press, Boulder, Colorado.

Delman, J. (1991). Agricultural extension in Renshou country China : A case study of bureaucratic intervention for agricultural innovation and change. Ph.D. Thesis, R&H Bulletin

No.28. AERDD, The University of Reading: Documentation Centre.

Dwarkinath, R. and Channegowda, M.B. (1974). Constraints to the transfer of new farm technology. In: Transfer Technology Systems and Constraints. Edited by D. Singh and J.S. Pandey. Nirbal Ke Balram Press, Kanpur, p.35.

Eponou, T. (1993). Partners in agricultural technology: Linking research and technology transfer to serve farmers. ISNAR Research Report No.1, The Hague: ISNAR.

Eponou, T. (1996). Partners in technology generation and transfer: Linkage between research and farmers' organization in three selected African countries. *ISNAR Research Report No. 9,* p. 8, The Hague, The Netherland.

Ernest, R.S. (1973). A study of communication utilization behaviour of small and big farmers and its implication to communication strategy. Ph.D. Thesis, I.A.R.I., New Delhi.

Esman, M.J. and Blaise, H.C. (1966). Institution building research-The guiding concepts. Inter University Research Program in Institution Building, University of Pittsburgh, Pittsburgh.

Farrington, J. (1997). Farmers' participation in agricultural research and extension: Lessons from last decades. Biotechnology and Development Monitor, *30* : 12–15.

Fernandez, F. (1982). Mechanism of inter institutional transfer of technology and information exchange. In : N.R. Usherwood (ed.). Transferring Technology for Small Scale Farming (ASA Special Publication No.41), Madison. WI: American Society of Agronomy.

Fischer, M.; David, S.; Farley, C. and Wortman, C. (1996). Applying farmer participatory research methods to planning agricultural research: Experiences from Eastern Africa. *J. Farming Systems Res. Extn., 6*(1) : 37–48.

Forest, L.B. (1990). 'We did it ourselves', Farmers' participation in research and extension. Development communication for agriculture. Edited by R.K. Samanta. B.R. Publishing Corporation, New Delhi.

Gadewar, A.U. and Ingle, P.O. (1993). Agricultural extension as a mechanism for technology transfer. *Maha. J. Extn. Edn., 12* : 359.

Ganorkar, P.L. and Bhugul, M.K. (1978). Adoption-decision behaviour of big and small farmers. *Agricultural Situation in India, 33*(4) : 223–227.

Ganorkar, P.L. and Khonde, S.R. (1979). Information transfer process - An evaluation study. *Agricultural Situation in India, 34*(2) : 699–703.

Guba, E.G. (1968). Development, diffusion and evaluation. In: Terry I. Sidell and Johne M. Kitchell (eds.). Knowledge Production and Utilization in Educational Administration. Fugene, Oregon: The Centre for Advanced Study of Educational Administration, University of Oregon, p. 235–250.

Gupta, A.K. (1991). Reconceptualising development and diffusion of technology for dry regions. Paper presented in International Conference on Extension Strategies for Minimising Risk in Rainfed Agriculture (April 6–8).

Gupta, H.C. (1974). Dairying as an instrument of change. *International Dairy Congress Report*, p.14, New Delhi.

Gupta, J. (1991). Progressive use of communication media by marine fishermen. *Jour. of Extension System, 7*(1) : 85–91.

Gupta, J. (1998). A study of the information management in dairy knowledge information system. Unpublished Ph.D. Thesis, J.V.C. Baraut, CCS University, Meerut.

Guttman, J.M. (1978). Interest groups and the demand for agricultural research. *J. Pol. Econ., 86*(31) :

Hafeez, A. and Subbarya, S.V. (1974). In : Pareek, U. and Rao, T.V. (eds.). Handbook of Psychological and Social Instrument, Samasthi, Baroda.

Hall, M. (1992). People are part of the pattern cares. The FAO Review, 24(6) : 31–35.

Hallriegal, D. and Slocum, J.M. (1974). Organizational climate measures research and contingencies. Acad. Mgt. J., *17*(2) : 255–280.

Hansara, B.S. (1996). A peep into the future extension. *J. Extn. Edn.*, *7* : 1386–1389.

Haverkort, A.W. and Roling, N.G. (1984). Six rural extension approaches. Wageningen : IAC, International Seminar on Strategies for Rural Extension.

Haverkort, B.; Fresco, L.O.F. and Engel, E. (1988). Strengthening farmers' capacity for technology development. *ILEIA*, Oct. 1988, *4*(3) : 3–7, ETC Foundation, Leusden, The Netherlands.

Hayami, Y. and Ruttan, V.W. (1971). Agricultural development: An international perspective. John Hopkins University Press, Baltimore.

Heaver, R. (1982). Bureaucratic politics and incentives in the management of rural development. World Bank Staff Working Paper No.537, Washington, DC.

Hewelock, R.G. (1971). Planning for innovation through dissemination and utilization of knowledge. Institute of Social Research, University, Michigan.

Huffman, W.E. and McNulty, M. (1985). Endogenous local public extension policy. *American J. Agril. Econ.*, *76*(4) : 761–768.

Huli, M.B. (1989). A study of managerial dynamics of performance of dairy plants. Ph.D. Thesis, NDRI Deemed University, Karnal.

Jain, N.C. (1970). Communication patterns and effectiveness of professionals performing linking roles in a research dissemination organization. Ph.D. Thesis, Michigan State University, USA.

Jain, T.C. (1985). Constraints on RE linkages in India. A World Bank UNDP Symposium edited by Cernea, Coultor and Russell.

Jakhar, B. (1993). Call for strengthened information coordination. *Pashudhan*, *8*(1) : 6.

Jones, G.E. (1990). Agricultural education: A system perspective. AERDD Bulletin, 29 : 3–6.

Kaimowitz, D. (1987). Research - technology transfer linkage. In: Report of the International Workshop on Agricultural Research and Management. International Service for National Agricultural Research held 7–12 Sept. The Hague, The Netherlands.

Kaimowitz, D. (1989). Linking research and technology transfer in the development of improved coffee technologies in Colombia. ISNAR Staff Notes, The Hague.

Kaimowitz, D. (1989). Placing agricultural research and technology transfer in one organization : Two experiences from Colombia. *Linkage Discussion Paper No.3, The Hague: ISNAR.*

Kaimowitz, D. (1990). Making the link: Agricultural research and technology transfer in developing countries. Boulder, Colorado: Westview Press.

Kaimowitz, D.; Snyder, M. and Engel, P. (1989). A conceptual framework for studying the links between agricultural research and technology transfer in developing countries. *ISNAR Linkages Theme Paper No.1, The Hague.*

Kaimowitz, D.; Snyder, M. and Engel, P. (1990). A conceptual framework for studying the links between agricultural research & TOT in developing countries. In making the link: Agril. Res. & Tech. Transfer Interface (Kainowitz, D., ed.) Westview Press, Boulder.

Kapoor, R.P. (1966). Relative effectiveness of information sources in the adoption process of some improved farm practices. In: Summaries of Extension by Post-graduate Students. PAU, Ludhiana.

Kaurani, M.D. (1995). Media support to agriculture. *Occasional Paper No.4. MANAGE,* Hyderabad, pp. 57-82.

Kessaba, A.M. (1989). Technology systems for resource-poor farmers, in Kessaba (ed.).

Khandwalla, P.N. (1977). Viable and effective organizational design of firms. *Acad. Mgmt. J., 16*(4) : 481-495.

Kolbe, D.A. (1974). Organizational Psychology, 2nd Ed. (N.J. Englewood Cliffs, ed.). Prentice Hall.

Krishnaraj, R. (1981). An analysis of organizational efficiency of milk producers' cooperative society - A systems approach. Ph.D. Thesis, NDRI Deemed University, Karnal.

Lawrence, R. and Lorsch, J.W. (1969). Organization and environment: Managing differentiation and integration. Richard D. Irwin Inc., Homewood, Illinois.

Lerner, D. and Schramm, W. (1967). Communication and change in developing countries. East West Centre Press, Honolulu, p. 294.

Lionberger, H.F. and Chang, C. (1970). Flow of farm information for modernising agriculture: The Taiwan System. Praeger Publishers, New York.

Maalouf, W.D. (1983). International experience in agricultural extension and its role in R.D. Paper presented on Regional Seminar on Extension & R.D. Strategies, Malaysia.

Mahipal and Kherde, R.L. (1989). Adoption of dairy innovations by medium and large farmers in relation to their socio personal and psychological characteristics. *Maharashtra J. Extn. Edn., 8* : 225-228.

Malik, J.S. (1993). Technology transfer model - An analysis of linkages. Unpublished Ph.D. Thesis, CCS HAU, Hisar.

Mamoria, C.B. (1966). Agricultural development problems of India. In : Rural reconstruction and community development. Kitab Mahal, Allahabad, p. 623.

Merill Sands, D. and Kaimouitz, D. (1990). The technology triangle: Linking farmers, technology transfer agents and researchers. *The Hague : ISNAR.*

Merrill Sands, D. and Kaimowitz, D. (1991). The technology triangle : Linking farmers, technology transfer agents and agricultural researchers. *ISNAR, The Hague.*

Merill Sands, D.; Ewell, P.; Biggs, S.; Bingen, R.J. McAllister, J. and Poats, S. (1990). Management of key institutional linkages in on farm client oriented research: Lesson from Nine National Agricultural Systems. *OFCOR Synthesis Paper No.1, The Hague: ISNAR.*

Mishra, S.K. (1994). Inaugural address at the National Seminar on Extension Education and Research and Development Linkage. Held at Conference Hall, IARI Library, New Delhi.

Mosher, A.T. (1978). An introduction to agricultural extension. New York : Agricultural Development Council, New York.

Murthy, A.S. (1969). Social and psychological correlates in predicting communication behaviour of farmers. Ph.D. Thesis, I.A.R.I., New Delhi.

Nagel, U.J. (1980). Institutionalization of knowledge flows : An analysis of the extension role of two agricultural universities in India. *Quarterly Journal of International Agriculture*. Special Volume Series No.30.

Nakkiran, S. (1968). The factors that contributed to the growth and success of the Tudialur Cooperative Agricultural Services Ltd., Tudialur, Coimbatore district, Madras state-A case study. M. Co op. Thesis, Sri Ramkrishna Mission Vidyalaya, Coimbatore.

Nawab, K. and Lawrence, L.D. (1995). Communication linkage among researchers, extension personnel and farmers in Pakistan. *J. Extn. Systems, 10*(1) : 37–45.

Pant, J.C. (1994). Suggested mechanism for strengthening research extension linkage in agriculture. Proceeding of the National Seminar on Extension Education and Research and Development Linkage held at Conference Hall, IARI Library, New Delhi.

Pareek, U.; Rao, T.V. and Pestonjee, D.M. (1981). Behavioural process in organizations. Oxford and IBH Publishing Co., New Delhi, Mumbai, Calcutta.

Patel, I.C. and Leagans, P.J. (1968). Some background and personal traits related to village level worker effectiveness. *Indian J. Extn. Edn.*, 4 : 3–4.

Patro, B.P. (1982). A study on the perception of management practices and their consequences in an agricultural research institutes. Ph.D. Thesis, NDRI Deemed University, Karnal.

Pestonjee, D.M. (1973). Organizational structure and job attitude. Calcutta, Minerva.

Pickering, D.C. (1985). Sustaining the continuum. In: Cernea, Coultar and Russel (eds.). Agricultural Extension by T&V. The Asian Experience, Washington, DC, World Bank.

Pineiro, M.E. (1989). Generation and transfer of technology for poor, small farmers. Technology Systems for Small Farmers-Issues and Options : Westview Press, Boulder, Sanfrancisco and London.

Pineiro, M. and Trigo, E. (1983). Social articulation and technical change. In : Technical Change and Social Conflict in Agriculture : Latin American Perspectives (M. ineiro and E. Trigo, eds.). Westview Press, Boulder, Colorado.

Prakasam, R. and Kshirsagar, S.S. (1971). Organizational climate in four banks. Report on a Survey *Prajnan, 8*(3):52-56.

Prasad, C. (1985). Linkage between agricultural research education and extension in India. A Country Paper for FAO of the United Nations: In Meeting held at Bangkok.

Prasad, C. (1988). Linkage between research and development systems: Concepts and implications. Paper presented at International Conference on Extension Strategies for Minimising Risk in Rainfed Agriculture (April 6–8).

Prasad, C. and Reddy, H.N.B. (1991). Extension strategy for minimizing risk in rainfed agriculture. In : Extension Strategies for Rainfed Agriculture. C. Prasad and P. Das (eds.). New Delhi : Indian Society of Extension Education.

Rahiman, O.A. (1991). Linkages between and among extension and other systems in rainfed rice farming. International

Conference on Extension Strategy for Minimising Risk in Rainfed Agriculture (April 6–8), Abstract.

Rao, S.V.N. (1992). Whey delay in farmers' adoption of dairy technologies. *Indian Dairyman, 44*(6) : 288–301.

Rao, S.V.N. and Sohal, T.S. (1980). Factors affecting job satisfaction of veterinary assistant surgeons of Vijayawada and Hyderabad Area (A.P.). *Indian J. Extn. Edn., 16*(3&4) : 74–77.

Reddy, H.N.B. and Singh, K.N. (1977). Analysis of communication pattern and procedures used by village level workers in Karnataka state. *Indian J. Extn. Edn., 13*(3&4) : 19–26.

Reddy, P.K. (1984). Analysis of information flow and communication linkages in transfer of dairy management practices - A system perspective. Unpublished Ph.D. Thesis, T.N.A.U., Coimbatore.

Rogers, E.M. and Pfizer, R.L. (1960). The adoption of irrigation by Ohio farmers. Research Bulletin 851. Wooster: Ohio Agricultural Experimental Station.

Rogers, E.M. and Shoemaker, F.F. (1971). Communication of innovations: A cross-cultural approach. New York Press.

Rogers, E.M. and Yost, M.D. (1960). Communication behaviour of country extension agents. Wooster, Ohio: Ohio Agricultural Experimental Station, Research Bulletin, p.850.

Rolling, N. (1989). Why farmers matter: The role of user participation in technology development and delivery. *The Hague: ISNAR.*

Roling, N. and Engel, P.G.H. (1992). IT from a knowledge system perspective: Concepts and issues. The edited proceedings of the European Seminar on Knowledge, Management and Information Technology. Agricultural University, Wageningen, The Netherlands.

Sah, A. (1996). A descriptive study of existing dairy farming practices and constraints in adoption of improved dairy practices among dairy farmers in Banka district (Bihar). M.Sc. Thesis, NDRI Deemed University, Karnal.

Samanta, R.K. (1990). Development communication for agriculture. B.R. Publishing Corporation, New Delhi.

Samanta, R.K. (1991). Agricultural knowledge transfer : Key for agricultural development. In: Agricultural Extension in Changing World Perspective, pp. 1-20.

Sandhu, M.S. (1967). A study of sources of information at different stages of adoption of selected improved agricultural practices by the cultivators of Ludhiana block. M.Sc. Thesis, P.A.U., Ludhiana.

Sanoria, Y.C. (1974). An analysis of communication patterns in farm innovation development and dissemination systems in Madhya Pradesh. Ph.D. Thesis, I.A.R.I., New Delhi.

Sanoria, Y.C. and Singh, K.N. (1978). Communication patterns of agricultural scientists: A system analysis. *Indian J. Extn. Edn., 14*(1&2) : 17–22.

Sarkar, A. (1981). A study of communication behaviour of dairy farmers in relation to scientific dairy farming practices of ICDP, Barasat (West Bengal). Ph.D. Thesis (NDRI), Kurukshetra University, Kurukshetra.

Sawant, G.K.; Thorat, S.S. and Phadtore, D.K. (1979). Information sources of the adopters of crossbreeding in cattle in Pune district of Maharashtra. *Indian J. Extn. Edn., 15*(1&2) : 75–76.

Seegers, S. (1990). The use of the training and visit system to link research and extension units for paddy production in Matara district, Sri Lanka. *ISNAR Linkages Discussion Paper No.8, The Hague.*

Seegers, S. and Kaimouitz, D.K. (1989). Relations between research and extension workers: The survey evidence. Linkage Discussion Paper No.2, The Hague: ISNAR.

Sharma, K. and Motilal (1971). Organizational climate and pupil achievement. *Rajasthan J. Edn., 7*(2) : 24–40.

Sharma, N.K. (1994). A comparative study of dairy development projects under different management systems in Haryana. Ph.D. Thesis, NDRI Deemed University, Karnal.

Sharma, N.K. and Rao, S.V.N. (1998). Beneficiaries' participation- An approach to measure the success and growth of dairy development organizations. *Indian J. Dairy Science, 50*(1): 12–17.

Sharma, P.K. (1982). Economic analysis of ICDP, Karnal. Ph.D. Thesis, NDRI, Karnal.

Shete, N.B. (1974). A study of communication behaviour of extension personnel of Maharashtra Agricultural Extension System. Ph.D. Thesis, I.A.R.I., New Delhi.

Sigman, V.A. and Swanson, B.E. (1982). Problem facing national agricultural extension in developing countries. Urbana, Illinois : INTERPAKS.

Sihag, S. and Grover, I. (1992). Communication pattern of farm women with reference to grape cultivation. *Maharashtra J. Extn. Edn., 11* : 28–34.

Sims, H. and Leonard, D. (1989). The political economy of the development and transfer of agricultural technology. *Linkages Theme Paper No.4, The Hague : ISNAR.*

Singh, A. and Gill, S.S. (1993). Review of adoption research studies in Indian Journal of Extension Education from 1980 to 1987. *Indian J. Extn. Edn., 29*(1&2).

Singh, B.K. (1970). Modernization and diffusion of innovations in a rural Appalachinan country - A general system analysis. Ph.D. Thesis, The University of Kentucky, Lexington Kentucky.

Singh, B.N. (1988). Recent researches in agricultural communication. In: Ostman, R.E. (ed.). Communication and Indian Agriculture, Sage Publications, New Delhi.

Singh, B.N. and Jha, P.N. (1965). Utilization of sources of information in relation to adoption of improved agricultural practices. *Indian J. Extn. Edn., 1*(1) : 34–42.

Singh, G. (1989). Communication system for human resource development of farmers in Haryana. Ph.D. Thesis, CCS HAU, Hisar.

Singh, K. (1998). Information dynamics in transfer of dairy production technologies in Kangra district of Himachal Pradesh. Unpublished Ph.D. Thesis, NDRI Deemed University, Karnal.

Singh, K.N. (1975). What research says about communicating with rural people. In: Communication and Rural Change. AMIC, Singapore.

Singh, P. (1994). Study on linkage between the State Agricultural University and State Department of Agriculture and Animal Husbandry. Unpublished Ph.D. Thesis, I.V.R.I., Izatnagar.

Singh, P. and Pastonjee, D.M. (1974). Supervisory behaviour and job satisfaction. *Indian J. Indus Relations, 9*(3) : 407-416.

Singh, R. (1989). A study of communication behaviour of dairy farmers regarding scientific dairy farming practices in ORP area of NDRI, Karnal. M.Sc. Thesis, NDRI Deemed University, Karnal.

Singh, S.K.; Singh, A.K. and Singh, D. (1991). Constraints operating in forward and backward linkages of monthly Workshop of T&V Extension, U.P. Society of Extension Education and Rural Development, C.S.A.U.A.&T., Kanpur, U.P.

Singh, S.N. (1967). Value orientation and adoption behaviour of Indian cultivators. Ph.D. Thesis, Ames, Iowa State University of Science and Technology.

Singh, S.N. (1974). Scale to measure achievement motivation. Cited from Handbook of Social and Psychological Instruments by V. Pareek and T.V. Rao.

Singh, S.P. (1980). Correlates of dairy modernization of small and marginal dairy farmers of ICDP Ludhiana. Ph.D. Thesis, NDRI, Karnal.

Singh, S.P. (1995). Operational Guide for IVLP. ICAR, New Delhi.

Singh, S.P.; Prasad, C. and Singh, D. (1991). Research development-farmer linkage for sustained higher productivity of rainfed agriculture. Extension Strategies for Rainfed Agriculture. Edited by Prasad and Das. Indian Soc. Extn. Edn., New Delhi.

Singh, T.P. (1997). Performance of non government organizations in dairy development among tribals of Ranchi (Bihar). M.Sc. Thesis, NDRI Deemed University, Karnal.

Singh, Y.P. (1965). A study of communication networks in sequential adoption and key communicators. Ph.D. Thesis, I.A.R.I., New Delhi.

Singh, Y.P. and Kumar, K. (1973). System of collecting and disseminating research information to extension workers and farmers. Paper presented at the National Symposium on Role of Extension in Agricultural University/Colleges/Institutions at I.A.R.I., New Delhi, p. 3.

Singh, Y.P. and Narwal, R.S. (1974). Audience analysis for using written words. Community Development and Panchayati Raj. Digest, *6*(1) : 22–28, NICD, Hyderabad.

Sinha, J.B.P. (1980). The authoritative leadership. *Indian J. Indus. Relations, 11*(3) : 381–389.

Sinha, P.R.R. and Prasad, R. (1966). Sources of information as related to adoption process of some improved farm practices. *J. Extn. Edn.,* 2(1) : 86–91.

Sivaraman, B. (1978). Address to the meeting of Vice Chancellors of Agricultural Universities (Mimeograph). Planning Commission, New Delhi.

Souder, W.E. (1980). Promoting an effective P&D/marketing interface. Research Management, *23*(4) : 10–15.

Sridhar, S. and Reddy, H.N.B. (1977). Communication patterns of extension personnel and the factors associated with them. M.Sc. Thesis, UAS, Hebbal, Bangalore.

Stephenson, T.E. (1963). Management of cooperative societies. Capetown, Auckland, William Heinemann Ltd., pp. 205– 232.

Supe, S.V. (1969). Factors relating to different degree of rationality in decision making among farmers. Ph.D. Thesis, I.A.R.I., New Delhi.

Suresh, S.V.; Jayaramaiah, K.M.; Shivamurthy, M. and Shivalingaiah, Y.N. (1995). Characteristics of marginal and small contact farmers. *J. Extn. Edn., 6*(2) : 1185 1187.

Swanson, B. and Peterson, W. (1991). Strengthening research-extension linkages to address the needs of resource-poor farmers in rainfed agriculture. In : Extension Strategies for Rainfed Agriculture. C. Prasad and P. Das (eds.). New Delhi: Indian Society of Extension Education.

Swanson, B.E.; Roling, N. and Jiggings, J. (1984). Extension strategies for technology utilization. In : Swanson, B.E. (ed.). Agricultural Extension : A Reference Manual. Rome: FAO, pp. 89–107.

Thurstone, L.L. (1946). The measurement of values. *Phychol. Rev., 61* : 47–58.

Trent, C.C. (1989). Strengthening the linkage between research and extension in the Ministry of Agriculture, The Gambia. Gambian Agricultural Research and Diversification Consultancy Report No. 54, Department of Agricultural Research. The Gambia : Ministry of Agriculture.

Trivedi, G. (1963). Measurement and analysis of socio-economic status of rural families: A study conducted in C.D. block, Kanjhala, Delhi State. Ph.D. Thesis, I.A.R.I., New Delhi.

Tyagi, K.C. and Sohal, T.S. (1984). Factors associated with adoption of dairy innovation. *Indian J. Extn. Edn., 20*(3&4) : 1–7.

Uphoff, N.; Boyton, D. and Whyte, W.F. (1983). Implications for Govt. policy. Higher Yielding Human Systems for Agriculture. Ithaca, NY : Cornell University Press.

Venkatesan, V. (1985). Mentioned in agriculture extension by training and visit : The Asian experience. The World Bank, Washington, D.C., USA.

Venkatesan, V. (1985). Policy and institutional issues in improving RE linkages in India. A World Bank & UNDP Symposium. Cernea, Coulter and Russel (eds.).

Verma, U. (1987). Analysis of communication pattern among information generating, information disseminating and information utilizing systems of Home Science in Haryana. Ph.D. Thesis, CCS HAU, Hisar.

Whyte, W.F. (1975). Organizing for agricultural development. New Brunswick, N.J. Transaction Books.

World Bank (1985). Agricultural research and extension : An evaluation of the World Bank's experience. World Bank, Washington.

Wuyts Fivawa, A. (1992). Management of intergroup linkages for agricultural technology system. Linkage Discussion Paper No.12, *The Hague: ISNAR*.

Yadava, J.P. (1971). Communication patterns and upward communication in C.D. Block agricultural administration. Ph.D. Thesis, I.A.R.I., New Delhi.

ANNEXURE - I

NATIONAL DAIRY RESEARCH INSTITUTE, KARNAL

The National Dairy Research Institute is a premier research Institute in Dairying dedicated to the cause of the dairy development in India. The scientific credibility, development of human resources and transfer of technologies have been the hall marks of the Institute programmes and activities. The Institute catalyses close interaction among scientists, students, farmers and dairy industry in such a way that each paves way for the other efficaciously : scientists, by undertaking teaching and research; students, by internalising the ever increasing knowledge; farmers, by not only adopting the new technologies in dairy farming, but also by providing the necessary feedback; and the industry, by posting problems encountered by it and seeking the expertise for their solutions. The Institute located at Karnal has two regional stations, one at Bangalore and the other at Kalyani, which provide further support in implementation of its mandate by carrying out research specific to their respective agro climatic zones.

NDRI is fully supported by the Indian Council of Agricultural Research and functions as one of the National Institutes under its aegis. The Institute also interacts with other national and international Institutes in Dairying and allied fields for exchange of information and advancing new knowledge, both in basic and applied fields of dairy science. The contributions of the Institute in conducting, collating and co-ordinating research in Dairying have received worldwide recognition.

The Imperial Institute of Animal Husbandry and Dairying established in Bangalore in 1923 earmarks the genesis of this

Institute. In 1936, it was expanded and renamed as Imperial Dairy Institute. Subsequently, in 1955, National Dairy Research Institute came into existence at Karnal at the location formerly called Central Cattle Breeding Farm, while the Institute at Bangalore was converted to function as a regional station for the South. Western Regional Station at Mumbai and Eastern Regional Station at Kalyani were established in 1962 and 1964, respectively. Western Regional Station at Mumbai was closed in 1984. The Institute was conferred the Deemed University status by the University Grants Commission in March, 1989.

With the concepts of national demonstration, adoption of villages, integrated technology transfer and appropriate technology generation, the Institute is closely associated in technology extension for resurgent Indian farmers. Dairy Extension Division, Krishi Vigyan Kendra and Farming Systems Research Project of the Institute have several well structured programmes for transfer of technology and dissemination of know-how developed at the Institute. These agencies link NDRI with the farming community of about 60 villages around Karnal. Krishi Vigyan Kendra (KVK) organises skill and production oriented short and long duration, on and off campus training programmes for practising farmers, men and women, and youth, whereas the Trainers' Training Centre (TTC) trains teachers of KVKs and Farmers' Training Centres, State Extension Functionaries and Dairy Plant Personnel. With the objective of providing necessary R&D support, consultancy services are offered to the dairy and cattle feed industry through the network of Industrial Consultancy Board set up at the Institute.

Objectives

During the past four decades, NDRI has experienced an enviable growth in the areas of research, training and technology transfer. It views its objectives for itself :

Studies on Dairy Production

Nutrient requirements of livestock for growth, milk production and reproduction, improvement in the nutritional quality of crop residues and agro industrial byproducts, biochemistry and physiology of growth, reproduction and lactation, livestock genetics, livestock breeding, management and housing of livestock to produce milk efficiently and economically under varying conditions.

Studies on Milk and Its Processing

Chemistry, microbiology and nutritive value of milk and its products, technology of milk processing and product manufacture including Indian dairy products, quality control, design and development of dairy equipment.

Studies on Dairy Economics, Management and Extension

Various costs that enter into production of milk, its processing and product manufacture under field, farm and factory conditions; dairy business management and extension of information; and development of dairy farming systems for different agro climatic conditions and varying types of land holdings and levels of operations.

Dairy Education

Organizing and conducting programmes both at the under graduate and post graduate levels in various branches of dairy science in an effort to provide manpower for not only milk production and processing programmes, but also for research and teaching needs of the country.

Training, Demonstration and Consultancy Services

Organizing short period training programmes for progressive farmers, field workers, plant operators, managers as well as arranging seminars, symposia for scientific personnel to provide current know-how, and helping in dissemination of knowledge, demonstrating package of practices using research findings for field applications and providing consultancy services to the industry and farming community.

Collaboration

Collaborating with national and international research and

training institutions in dairying and allied fields, for exchange of information and advancing new knowledge, both in basic and applied fields of dairy science.

ORGANIZATIONAL STRUCTURE

The Institute operates through five main bodies which are responsible for policy matters and decision making in the field of research, education and training, extension education and administration. They are : Board of Management, Executive Council, Research Advisory Committee, Academic Council and Extension Council.

The highest policy making body is the Board of Management. The Director, NDRI is the Chairman of this Board. The Executive Council is the main task implementing body on administrative matters. The Research Advisory Committee is responsible for all round progress of research at the Institute and its application. The Academic Council is responsible for all issues relating to the education and training. The Academic Council, in turn, is supported by (*i*) a number of Standing Committees, (*ii*) the post-graduate faculty, and (*iii*) the Board of Studies in the respective disciplines. The Extension Council is responsible for extension programmes.

The Research, Education and Extension activities of the Institute are managed by the Director and the Joint Directors through Scientific, Technical, Auxiliary, Administrative and Supporting Staff. The Director is the Administrative Head of the Institute and its Regional Stations. The Joint Directors (Academic and Research) in addition to helping the Director in administration, are responsible to coordinate education and research activities of various Divisions and Regional Stations, respectively. Each of the Regional Stations is administered through a Head located at the Station. The scientific and teaching work at the main station is conducted through ten subject matter Divisions.

ANNEXURE - II

CHAUDHARY CHARAN SINGH HARYANA AGRICULTURAL UNIVERSITY, HISAR

The actual role performance of agricultural universities in rural development is concerned, different agricultural universities have come out with varying outcome of their roles and performance depending upon the state's interests, financial, and other limitations. The actual role played by some of the leading agricultural universities in the country, including Chaudhary Charan Singh Haryana Agricultural University, Hisar in rural development may be stated as follows :

- Imparting instructions in the various subjects of Agriculture, Veterinary Sciences, Animal Sciences, Home Sciences, and Basic Sciences so as to provide trained manpower for the scientific development of agriculture, livestock, farm and home and allied rural sectors.
- Generating knowledge and development of appropriate technology through research for providing solutions to the problems confronted in optimising agricultural and livestock production, and improvement of rural households for a better life.
- Undertaking location specific agricultural research with the establishment of strong multi disciplinary regional research stations so as to meet the specific research requirements of different agro climatic situations to develop need-based technology for the farmers.
- Transferring the applicable and useful research findings and improved technologies in agriculture and allied

fields to the concerned clientele quickly and efficiently. This includes :

- advisory work and feedback of farmers' problems to the concerned subject matter departments of the university for finding solutions for onward transmission to them;
- organizing and conducting training programmes of short and long durations for different categories of clientele comprising farmers, farm women, rural youths, subject matter specialists, field functionaries of agriculture, animal husbandry and development departments, school teachers, defence personnel and others like Bank Officers, Indian Administrative Services and State Services Probationers;
- information communication through various mass media utilizing a variety of channels suited for the purpose;
- publication of popular and extension literature in the form of leaflets, handouts, magazines in local language for education of farmers on improved practices and technologies;
- production of audio visual material for utilization by the farmers and field functionaries for learning improved technology;
- organization of Kisan Melas, Farm Darshans, etc., for facilitating the exposure of farmers to improve methods of farming by arranging their visits to the university farms agricultural exhibition and providing them opportunities for having discussions of their problems with the subject matter specialists of the university;
- organizing workshops at the university campus for the agricultural and animal husbandry officers of the State Departments to keep abreast of the latest advances in the profession, developing programme

of work for implementation in the field based on the package of practices finalised in these workshops; and

- establishing an efficient and effective linkage between the University and Development Departments of the State so as to ensure maximum functional collaboration between the field functionaries of the State Government and the Scientists of the University for successful implementation of the extension programmes at the grassroots level.

It may not be out of context to mention here that HAU, with its unique model of 'Extension Education System' could achieve these objectives to an appreciable degree.

No doubt, the above mentioned functions of the agricultural universities have gone a long way in increasing agricultural production in several states and even most of the developed countries have applauded India's efforts. There has also been an appreciable socio economic transformation leading to the improvement in the living standard of rural people in several states.

EXTENSION EDUCATION

The Directorate of Extension Education of the University has the most widespread, strong and unique network in the State. Its extension education activities are conducted through three wings of Directorate: Farm Advisory Service, Farm Training Service and Farm Information Service.

Farm Advisory Service

The Farm Advisory Service is the major wing and field arm of the Directorate, covering the entire state of Haryana through its FAS centres, generally known as Krishi Gyan Kendras (KGKs) and Krishi Vigyan Kendras (KVKs). The scientists working at these centres have direct contact with farmers and render necessary advice at their doorsteps.

The scientists maintain active liaison with field functionaries of Department of Agriculture, Animal Husbandry and various development agencies directly or indirectly involved with rural development, rural youth, rural institutions, agricultural business complex, government departments, administrators, legislators, local leaders and other rural people and agencies having a bearing on agricultural and rural development. Various extension activities like trainings, demonstrations, workshops, field days, gyan diwas, district level training camps, symposia, exhibitions, crop campaigns, krishi melas, farm darshan, animal camps, radio talks, T.V. programmes, summer institutes and extension publications are organized by Farm Advisory Service.

Farm Training Service

The training unit organises about 130 to 150 trainings per year, which are of 1 to 90 days duration and participated by 4000–5000 persons. These trainings, mostly institutional, benefit farmers, urban and rural women, unemployed youth, livestock owners, field functionaries, Govt. officers, administrators, bankers, NGOs and voluntary organizations. Some off-campus courses are also conducted by farm training service.

These trainings relate to agriculture, animal and home sciences. Trainings in agriculture pertain to production and protection technology of field crops, vegetables and fruits. Specific courses on integrated pest management, integrated nutrient management, use of bio-fertilisers, safe storage of food grains, nursery raising for fruits and vegetables, surveillance and forecasting of pests and diseases, are also run. For the transfer of technology, trainings on use of A.V. aids; maintenance and operation of A.V. equipment; extension methods and techniques; rapid rural appraisal and participatory rural appraisal are also conducted. Animal Science courses include poultry farming; piggery, poultry, dairy farming, milk and milk products; quality meat production; quality evaluation of eggs and meat, etc. Important courses related to home sciences are, nutrition education and promotion; candle, toys, dolls making, detergent making, knitting and garment making, banking, etc.

Several of the training programmes are sponsored by the Govt. of India, Ministry of Agriculture, Ministry of Non Conventional Energy Sources, Ministry of Human Resource Development, Govt. of Haryana and nationalised banks. The Advance Training Centre established by Govt. of India, Ministry of Agriculture caters to the training needs of senior and middle level officers of oilseeds producing states of the country. From this year, IATTE has been entrusted with the responsibility of organizing trainings for rural women under the Central sector scheme on 'Women in Agriculture', being implemented as a pilot project in the district of Hisar. The Govt. of India, Ministry of Non conventional Energy Sources has also sanctioned a Regional Biogas Development and Training Centre being established at Kisan Ashram to train turn key workers, masons, govt. staff, biogas users, panchayat workers, etc. in biogas development.

Much sought after courses by FTS include mushroom production, bee-keeping, floriculture, fruit and vegetables preservation, piggery and poultry farming, fish farming, garments construction, bakery, embroidery, etc.

Farm Information and Communication Service

The responsibility for planning, organizing and coordinating Extension Education activities/programmes at the university level lies with the Directorate of Extension Education. It aims at transfer of technology to the farmers and farm women through its three wings namely: Farm Advisory Service, Farm Training Service and Farm Information and Communication Service (FICS).

The objective of Farm Information and Communication Service is to provide mass media and audio-visual aids support to the other two wings of the Directorate as well as to the constituent Colleges/Departments/Krishi Gyan Kendras (KGKs)/Krishi Vigyan Kendras (KVKs) of the University. To achieve this objective, this section performs multifarious responsibilities ranging from preparing exhibits, models, display boards and charts, providing audio visual aids support and repair of audio-visual equipments, publishing of bulletins, leaflets, folders, posters and handbills, arranging and organizing agro-industries exhibitions at various occasions and participation in International, National and State level exhibitions/ Agriculture Fairs organized by various agencies.

ANNEXURE - III

STATE DEPARTMENT OF ANIMAL HUSBANDRY, KARNAL

Under SDAH, the Project of Intensive Cattle Development (ICDP) was launched in the year 1964 as a part of the Special Development Programmes initiated during the later half of the Third Five Year Plan. The approximate cost estimated to establish an ICDP was Rs. 1.20 lakhs for a period of five years. These projects were specially designed to cover one lakh breedable cow and she buffalo population with the network of a Central Semen Bank, four Regional Artificial Insemination Centres and 100 Stockmen Centres, for creating a discernible impact on milk production. It was envisaged at the time to locate these projects in the breeding tracts of indigenous breeds of cattle and buffalos, and in the milk shed areas of large milk plants with the idea of enabling these plants to collect and process milk up to their installed capacities. Currently, all the districts in Haryana are covered by ICDPs.

Objectives of the SDAH/ICDP

(*i*) Providing facility for controlled breeding.

(*ii*) Providing health care.

(*iii*) Arranging for marketing of milk.

(*iv*) Supplying feeds and popularising fodder cultivation.

Activities of the SDAH/ICDP

(*i*) **Artificial Insemination :** Each Civil Veterinary Dispensary (CVD) covers around 3000 breedable animals. Each CVD covers an area within a radius of two km.

Stockman also renders services like A.I., pregnancy diagnosis, prevention of diseases, first-aid treatment, etc.

(*ii*) **Health Care :** One Veterinary Surgeon (VS) is deputed for each Civil Veterinary Hospital (CVH) and Hospital-cum-Breeding Centre (HCBC). Each CVH covers 4-5 CVDs. It caters to the health care need of all types of livestock in that area. At Karnal ICDP, one Clinical Diagnostic Laboratory (CDL) is also under operation.

(*iii*) **Prophylactic Vaccination :** CVDs take up regular vaccination programme, specially against HS, BQ, GTV, TCV and FMD in cattle, buffaloes and against Ranikhet and fowl pox diseases in case of poultry birds.

(*iv*) **Popularising Fodder Cultivation :** ICDPs supply improved fodder seeds at a subsidised rate (50%) to the farmers. Minikits (mainly oat, berseem, guar and jowar) are also distributed by the State Agricultural Department through ICDPs.

(*v*) **Semen Bank :** Each ICDP has got a Semen Bank with liquid nitrogen plant. This plant at Karnal was not functioning during the study period. ICDP, Karnal mainly supplies liquid semen to all the CVHs/CVDs on regular basis. Whereas, ICDP, Gurgaon supplies frozen semen to its CVHs/CVDs for A.I. purposes in cattle and buffaloes.

ANNEXURE - IV

VARIABLES PERTAINING TO SCIENTIFIC AND EXTENSION PERSONNEL

A. PERSONAL TRAITS

1. Name :
2. Designation :

(*a*) Basic Pay :

(*b*) Cadre :

3. Age :
4. Education : High School with Diploma
   Intermediate
   B.Sc. (Ag.)
   B.V.Sc. & A.H.
   M.Sc. (Ag.)
   M.V.Sc.
   Ph.D.
5. Professional Experience :

Sr.No.	*Post held*	*Experience (Years)*	*Organization*

6. Training Received : Yes/No

Sr.No.	*Field/Area of Training*	*Duration*	*Organizing Institute*

7. Family :

(*i*) Type of family : Nuclear/Joint

(*ii*) Family background : (*a*) Rural/Urban

(*b*) Agri./Non Agriculture

(*iii*) Family Size :

(Total No. of family members)

B. PSYCHOLOGICAL/BEHAVIOURAL TRAITS/ VARIABLES

1. Attitude

	Your response (kindly √ mark)		
(*a*) With respect to your work in this organization :			
(*i*) Your interest in the work	(*a*) Real enjoyment in my work	(*b*) My work is usually interesting	(*c*) Other jobs are more interesting than my work.
(*ii*) Your idea about your present work	(*a*) My work is best and I would not change it	(*b*) I am forced to do this present work	(*c*) I definitely dislike my work
(*iii*) Creativity in your work.	(*a*) Gives opportunity to express myself completely.	(*b*) Does not give any chance to improve myself	(*c*) The work does not require my creative ability at all.
(*b*) With respect to your working condition :			
(*i*) Your idea about the present working condition.	(*a*) Working conditions are of high order/class	(*b*) Working conditions on many aspects can be improved.	(*c*) Condition not better than those in other places.
(*ii*) Your satisfaction with your place of work.	(*a*) This place is best possible place to work.	(*b*) Much though should be given to the working conditions.	(*c*) Not all suitable for working under the present condition.
(*c*) With respect to co-workers in your organization/department :			
(*i*) In terms of courtesy of the employee.	(*a*) Co-workers are courteous to each other.	(*b*) Neither overbearing nor courteous.	c) Little friendship exists among workers here.
(*ii*) In terms of cooperation.	(*a*) We always help in each other's work.	(*b*) We are friendly but everybody minds his own business.	(*c*) Eventhough we are friendly one need not help other.

(Contd...)

(*d*) With respect to your supervisory/immediate boss :			
(*i*) Supervisors/Boss's idea of their job.	(*a*) Know their job very well.	(*b*) Supervisors/Boss need some more training.	(*c*) Their/his knowledge of the work is much less.
(*ii*) Regarding his helping nature.	(*a*) Help us all without any partiality.	(*b*) Supervisors help only a few.	(*c*) Never help or advise us.
(*iii*) In case of any failures or loss in the work/duty, the supervisor/boss.	(*a*) Share have responsibility position by being.	(*b*) Keep himself unconcerned.	(*c*) Tries to shift his responsibility.
(*iv*) In terms of his capabilities.	(*a*) They have earned their position by being good and capable.	(*b*) They could have been more capable.	(*c*) They pretend to be capable man.
(*e*) With respect to the organization as a whole where you are working :			
(*i*) Your enthusiasm.	(*a*) Feel enthusiastic to work for this organization.	(*b*) This organization is good enough to work for.	(*c*) I would get out of this organization as soon as I will find some other work.
(*f*) With respect to the management of this organization :			
(*i*) Management's interest in the employees.	(*a*) Management's assistance is highly satisfactory.	(*b*) Management is interested only in getting the work done.	(*c*) Management treats us as a machine in organizational work.
(*ii*) Authority's fairness in job distribution.	(*a*) Management is fair and always look out for the right person for right job.	(*b*) Management does not give job according to capacity and ability.	(*c*) The work does not at all go to the right person here.
(*iii*) In terms of your job security.	(*a*) My job remains intact irr-espective of my performance.	(*b*) Authority can discharge me any time in their own interest.	(*c*) The job is highly insecure because of the authorities.

2. Achievement Motivation

Kindly tick (√) out the most suitable option as perceived by you against each statement in relation to your job.

S.No.	*Statements*	*Response*				
1.	Success bring further determination	Strongly agree	Agree	Un-decided	Dis-agree	Strongly disagree
2.	How true is it to say your efforts are directed towards success?	Quite natural	Not very true	Not sure	Fairly true	Quite true
3.	How often do you seek opportunity to excell?	Hardly ever	Seldom	About half the time	Fre-quently	Nearly always
4.	Would you hesitate to undertake some assignment?	Hardly ever	Seldom	About half the time	Fre-quently	Nearly always
5.	In how many spheres that might lead to your feelings?	Most	Many	Some	Very few	Few
6.	How many situa-tions do you think that you will succeed in doing so as well as you can?	Most	Many	Some	Very few	Few

3. Value Orientation

Please state the degree of your agreement or disagreement on the given 4 point scale against each statement.

Sr.No.	*Statement*	*Response*			
		Strongly Agree	*Agree*	*Disagree*	*Strongly Disagree*
1.	An employee learn many things from the happening and experience of his working organization.				
2.	One's experience is not as better as collective experience of a group of staff.				

(Contd...)

Sr. No.	Statement	Response: Strongly Agree	Agree	Disagree	Strongly Disagree
3.	An employee who has been something worked in his organization need not worry about taking any additional information from outside his organization.				
4.	An employee who does not believe in consulting others can do a better job.				
5.	One can satisfy all his requirements out of the local resources (within organization) available to him.				
6.	An employee can save himself with many decisions and difficulties during his service, if he believes in taking and following the advice of other colleagues.				
7.	Many times an employee ought to know about happenings outside of his organization and such happenings may be of great advantage to him.				
8.	It is a sign of weakness and impotency when an employee relies on other's opinions for making his decisions.				
9.	At present, when transport and other communication facilities are developing, an employee should know more about things happenings outside his organization.				
10.	Action should be undertaken only after consulting others.				

4. Job Satisfaction

Some jobs are more interesting and satisfying than others. We want to know how you feel about your present job. This blank contains 18 statements about jobs. You are requested to place a tick mark (√) in the square depending upon your degree of agreement or disagreement for such statement. There are no right or wrong answers. We would like your honest opinion on each one of these statements.

Sr. No.	*Statement*	*Strongly Agree*	*Agree*	*Un-decided*	*Disagree*	*Strongly Disagree*
1.	My job is like a hobby to me	5	4	3	2	1
2.	My job is usually interesting enough to keep me from getting bored.	5	4	3	2	1
3.	It seems that my co-workers are more interested in their jobs.	1	2	3	4	5
4.	I consider my job rather unpleasant.	1	2	3	4	5
5.	I enjoy my work more than every my leisure time.	5	4	3	2	1
6.	I am often bored with my job.	1	2	3	4	5
7.	I feel fairly well satisfied with my present job.	5	4	3	2	1
8.	Most of the time I have to force myself to go to work.	1	2	3	4	5
9.	I am satisfied with my job for the time being.	5	4	3	2	1
10.	I feel that my present job is no more interesting than others I could get.	1	2	3	4	5
11.	I definitely dislike my work.	1	2	3	4	5
12.	I feel that I am happier in my work than most other people.	5	4	3	2	1
13.	Most days, I am enthusiastic about my work.	5	4	3	2	1
14.	Each day of work seems like it will never end.	1	2	3	4	5
15.	I like my job better than the average worker does.	5	4	3	2	1
16.	My job is pretty uninteresting.	1	2	3	4	5
17.	I feel real enjoyment in my work.	5	4	3	2	1
18.	I am disappointed that I ever took this job.	1	2	3	4	5

5. Personnel's Morale

Please tick (√) three statements as you feel suitable from each section. There are four sections and you have to check twelve statements in total.

Section - I

1. Employees here get a fair deal.
2. Staff welfare is considered as most important here.
3. Ability of workers are respected more here than in any other sister organization.
4. Good work is praised sometimes and sometimes not.
5. Partiality is considered against the policy of this Institute.
6. Employees cannot raise their voice for their own welfare.
7. Ordinary workers are not considered as human beings.

Section - II

1. Only able persons are appointed as superior officers here.
2. Officers here consider the welfare of staff as their own.
3. Some workers are more capable than their superiors.
4. Superiors understand the difficulties of everybody.
5. For authorities here, a good man is one, who is a good worker.
6. Superiors want their own welfare, not that of subordinates.
7. There are no such qualities in the superiors here for which they may be praised.

Section - III

1. Everybody is consulted for the welfare of the Institute.

2. Employees are encouraged to suggest new ideas about the work.
3. Employees are bound to work in a particular method, so there is no enthusiasm for work.
4. Employees are free to apply new technique (method) of work according to their own will.
5. In other sister organizations, employees are getting more opportunity to show their abilities and to utilise their past experiences.
6. Workers are never consulted about the work.
7. It is believed here that the progress of the Institute depends only on its superior officials.

Section - IV

1. Employees are always willing to do everything for the Institute.
2. The future of the organization and the future of the employees is the same.
3. My Institute get respects from every quarters.
4. Employees of this Institute are better-off than employees of any other similar institutions.
5. I will not advice any of my relatives and friends to work here.
6. The progress of the Institute does not provide any benefit to the employees.
7. Most of the people work here under the conditions of helplessness or fear.

6. Perception of Management (PMS) Scale

Please put (√) mark against each statement as you feel appropriate in relation to your Institute under given column.

Sr. No.	*Statement*	*Strongly Agree*	*Agree*	*Un-decided*	*Disagree*	*Strongly Disagree*
1.	Planning and formulation of research/extension projects are as per the need of farmers.					
2.	Decision made with respect to research/extension activities in this Institute are executed without much delay.					
3.	Officers are familiar with the various rules, regulations of the organization.					
4.	Rules are not rigid but flexible in this Institute.					
5.	Some personnel/staff are overloaded and some are under work.					
6.	Sentiments and feelings of employees are cured while assigning certain duties to them.					
7.	Maximum responsibility is given to junior levels, but withheld the authority at senior levels.					
8.	Authorities and powers are centralised in this Institute.					
9.	Selection of personnel to various posts of this Institute is impartial.					
10.	Punishment/reward in this Institute is based on caste, creed and region.					
11.	Authorities praise good workers in this Institute.					
12.	Grievances are not settled unless they are voiced through other quarters.					
13.	The relationship between different categories of staff is cordial.					

(Contd...)

Sr. No.	*Statement*	*Strongly Agree*	*Agree*	*Un-decided*	*Disagree*	*Strongly Disagree*
14.	Jealousy and leg pulling is prevalent in this Institute.					
15.	Farmers programme is not effective because of poor co-ordination.					
16.	A subordinate does not feel free to report his problems to superior.					
17.	The communication network between farmers and Institute is adequate.					
18.	There is no misappropriation of funds in this Institute.					

C. ORGANIZATIONAL VARIABLES

1. Goals of the Organization/Department

Which of the following goals or objectives guide day do day as well as long term decisions of your organization/ department. Your organization will be having its own priorities on the goals listed below. So kindly assign scores depending upon the importance each goal has got. Please also see that total score equals 100.

		Scores
1.	Department is strongly committed to improve the indigenous cattle and buffalo breeds in the area under its operation.	______
2.	Department is strongly committed to bring down the mortality rate in cattle and buffalo by providing better veterinary facilities.	______
3.	Department is concerned to improve the milk yield of cows and buffaloes.	______
4.	Organization is concerned with only basic and applied researches in animal science.	______
5.	Department also believes that the cattle and buffalo owners should get required inputs in time and at appropriate price.	______

(Contd...)

6. Providing technical inputs and extension services to farmers to increase milk production is the main concern of the department. ______
7. Department believes that the farmers should be educated about management aspects like breeding, feeding, disease prevention measures. ______
8. Organization is committed to generate employment opportunities for the educated unemployed youths traditionally engaged in the profession of cattle and buffalo rearing. ______
9. Fodder development is one of the objectives of the organization. ______
10. Organization aims at improving the socio economic status of weaker sections through dairying.

Total = 100

2. External Environment of the Organization

Kindly give your response (√) about following external factors constraining the research/extension activities by your organization/department.

Sr.No.	External Environment	*Constraints the Activity*			
		Yes	*No*	*Uncertain*	*Not Relevant*
1.	Government policy				
2.	Farmers organization/group				
3.	Foreign agencies				
4.	Private organizations				
5.	Any other (Please specify)				

3. Organizational Climate

We are placing before you some of the dimensions of organization climate. Kindly place an (A) above the number that indicates your assessment of the organization's current positions on that dimension and an (I) above the number that indicates your choice of where the organization should ideally be on this dimension.

1. **Conformity :** The feeling that there are many externally imposed constraints in the organization; the degree to which members feels that there are many roles, procedures, policies, and practices to which they have to conform rather than being able to do this work as they see fit.		
Conformity is not characteristic of this organization.	1 2 3 4 5 6 7 8 9 10	Conformity is very characteristic of this organization.
2. **Responsibility :** Number of the organizations are given personal responsibility to achieve their part of the organization's goals; the degree to which members feel that they can make decisions and solve problems without checking with superiors each step of the way.		
No responsibility is given in organization.	1 2 3 4 5 6 7 8 9 10	There is a great emphasis on personal responsibility in the organization.
3. **Standards :** The emphasis of the organization on quality, performance and outstanding production including the degree to which the members feels the organization in setting challenging goals for itself and communicating these goals commitments to members.		
Standards are very low on non-existent in the organization.	1 2 3 4 5 6 7 8 9 10	High challenging standards are set in the organization.
4. **Rewards :** The degree to which members feel that they are being recognised and rewarded for good work rather than being ignored, criticised, or punished when something goes wrong.		
Members are ignored, punished or criticised.	1 2 3 4 5 6 7 8 9 10	Members are recognised and rewarded positively.
5. **Organizational Clarity :** The feeling among members that thins are well organized and goals are clearly defined rather than being disorderly confused, or chaotic.		
The organization is disorderly, confused and chotic.	1 2 3 4 5 6 7 8 9 10	The organization is well organized with clearly defined goals.
6. **Warmth and Support :** The feeling that friendliness is a valued norm in the organization; that members trust one another and offer support to one another. The feeling that good relationship prevail in the work environment.		
There is no warmth and support in the organization.	1 2 3 4 5 6 7 8 9 10	Warmth and support are very characteristic of the organization.

7. **Leadership** : The willingness of organization members to accept leadership and direction from qualified others. As needs for leadership arise members feel free to take leadership roles and are rewarded for successful leadership. Leadership is based on expertise. The organization is not dominated by, or dependent on, one or two individuals.

Leadership is not rewarded; members are dominated or dependent and resist leadership attempts.	1 2 3 4 5 6 7 8 9 10	Members accept and rewarded leadership based on expertise.

ANNEXURE - V

VARIABLES PERTAINING TO DAIRY FARMERS

Name of Village : Sr. No. _______

Adopted by : Date _________

Name of Respondent :

A. SOCIO-PERSONAL VARIABLES

1. Age :
2. Education (1-8) : Illiterate/Can Read Only/Can Read and Write/
   (1) (2) (3)
   Primary/Middle/High School/Intermediate/
   (4) (5) (6) (7)
   Graduate/Post graduate
   (8) (9)
3. Family Education Status

Sr.No.	*Name of Family Members*	*Educational Level*								
		1	2	3	4	5	6	7	8	9
1.	Self									
2.	Wife									
3.	Father									
4.	Mother									
5.	Son/Daughter (above 6 years)									
	_________									
	_________									
	_________									

(Contd...)

Sr.No.	Name of Family Members	Educational Level								
		1	2	3	4	5	6	7	8	9

4. Family Type	:	Nuclear/Joint	
5. Family Size	:		
6. Occupation	:	Primary/Main -	Agri./Dairying/Business/ Contract Labour/Govt./Pvt. Service
		Secondary -	- do -
7. Caste	:		SC/ST/OBC/FC

B. SOCIO-ECONOMIC VARIABLES

1. Land Holding (in acre).
2. Herd Size.

Breed	In Milk	Dry	Heifer	Calves	Bullocks
Desi					
Crossbred					
Buffalo					
Total					

3. Milk Production/Consumption/Sale (litres/day) : (a day prior to interview).

Category	Production	Consumption	Sale
Cow			
Buffalo			
Total			

C. COMMUNICATION VARIABLES

1. Social Participation

Have you been associated with the following organization?

Sr. No.	Organization	Member	Office Bearer
1.	Gram Panchayat		
2.	Panchayat Samiti		

Sr. No.	Organization	Member	Office Bearer
3.	Zila Parishad		
4.	Agril. Coop. Society		
5.	Milk Coop. Society		
6.	Religious Committee		
7.	Political Organization		
8.	Rural Youth Club		
9.	Mahila Mandal		
10.	Any other (specify)		

2. Extension Contact

(*a*) How often (No.) during the last year you met the following in connection with dairy farming?

(*b*) How often (No.) during the last year you invited following in the same connection.

Sr.No.	Person	Frequency	
		(*a*)	(*b*)
1.	Veterinary doctor of vety. dispensary		
2.	Stockman of SMC		
3.	Expert from KGK		
4.	Expert from NDRI		
5.	Experts from Dairy Cooperative		
6.	Experts from Dairy Development Department		
7.	Others, if any		
	__________		
	__________		

3. Mass Media Exposure

Kindly give your degree of exposure to the following mass-media in relation to dairy farming.

Sr.No.	Media	Regularly (3)	Occasionally (2)	Never (1)
1.	Radio			
2.	T.V.			
3.	Newspaper			
4.	Dairy Samachar/Newsletter			
5.	Farm Magazine			
6.	Dairy Mela			
7.	Exhibition			
8.	Film Show			
9.	Meeting			
10.	Others, if any			
	____________			
	____________			

D. PSYCHOLOGICAL VARIABLES

1. Risk Preference

Sr. No.	Statement	Strongly Agree	Agree	Un-decided	Disagree	Strongly Disagree
1.	A farmer rather should take more of a chance in making a big profit than to be content with a smaller but less risky profit.	7	5	4	3	1
2.	A farmer who is willing to take greater risk than the average farmer usually does better financially.	7	5	4	3	1
3.	A farmer should grow large number of crops to avoid greater risks involved in growing one or two crops.	1	3	4	5	7

(Contd...)

Sr. No.	Statement	Strongly Agree	Agree	Un-decided	Disagree	Strongly Disagree
4.	It is better for a farmer not to try new farming methods unless most other farmers have used them with success.	1	3	4	5	7
5.	Trying entirely new method in farming by a farmer involves risks but it is worth of it.	7	5	4	3	1
6.	It is good for a farmer to take risk when he knows his chances of success are fairly high.	7	5	4	3	1

2. Cosmopolite - Localite

Sr. No.	Statement	Strongly Agree	Agree	Disagree	Strongly Disagree
1.	These days, when communication has so much advanced, farmers should know more of outside life.	4	3	2	1
2.	A farmer can learn everything from experience of his own village.	1	2	3	4
3.	He who does not consult others, can act better.	1	2	3	4
4.	A man can escape numerous troubles and worries, if he consults friends and neighbourers.	4	3	2	1
5.	A farmer can fulfil all his needs with the help of his villagers.	1	2	3	4
6.	Many folks thinks that a farmer ought to know are not only confined in his village but are also alike in other village.	4	3	2	1

E. SOURCE PERCEPTION

From whom you could know about the management, activities (Research and Extension) of the Institution :

(*i*) Self experience

(*ii*) Extension staff

(*iii*) Relative working

(*iv*) Friends

(*v*) Any other

F. PERCEPTION OF MANAGEMENT

Please put (√) mark against each statement as you feel appropriate in relation to your Institution under given column.

Sr. No.	*Statement*	*Strongly Agree*	*Agree*	*Un-decided*	*Disagree*	*Strongly Disagree*
1	*2*	*3*	*4*	*5*	*6*	*7*
1.	Planning and formulation of research projects are as per the need of farmers.					
2.	Decisions made in this Institute are executed without much delay.					
3.	Officers are familiar with the various rules, regulations of the the organization.					
4.	Rules are not rigid but flexible in this Institute.					
5.	Some workers are overloaded and some are under work.					
6.	Sentiments and feelings of employees are cared while assigning certain duties to them.					
7.	Maximum responsibility is given to junior levels, but withheld the authority at senior levels.					
8.	Authorities and powers are centralised in this Institute.					
9.	Selection of personnel to various posts of this Institute is impartial.					
10.	Punishment/reward in this Institute is based on caste, creed and region.					

(Contd...)

1	2	3	4	5	6	7
11.	Authorities praise good workers in this Institute.					
12.	Grievances are not settled unless they are voiced through other quarters.					
13.	The relationship between different categories of staff is cordial.					
14.	Jealousy and leg pulling is prevalent in this Institute.					
15.	Farmers programme is not effective because of poor co-ordination.					
16.	A subordinate does not feel free to report his problems to superior.					
17.	The communication network between farmers and the Institute is adequate.					
18.	There is no misappropriation of funds in this Institute.					

G. KNOWLEDGE ABOUT SCIENTIFIC IMPROVED DAIRY FARMING PRACTICES

1. What are the exotic breeds known to you?
2. After how many days a normal cow comes into regular heat?

   (*a*) 18 to 24 days.

   (*b*) 19 to 22 days.

   (*c*) 15 to 20 days.

   (*d*) No answer.
3. What is the right time of artificial insemination (AI) when a cow is in heat?

   (*a*) Between 12 and 18 hours.

   (*b*) Between 19 and 24 hours.

(*c*) More than 24 hours.

(*d*) Any other/No answer.

4. After how many days of calving, a normal cow/buffalo should be inseminated, if it comes in regular and normal heat?

(*a*) Between 60 to 90 days.

(*b*) Between 91 and 120 days.

(*c*) More than 120 days.

(*d*) Any other/No answer.

5. At what age, a crossbred heifer with average body weight should be inseminated?

(*a*) Between 15 and 18 months.

(*b*) Between 19 and 22 months.

(*c*) More than 22 months.

(*d*) Any other/No answer.

6. After how many days of AI, a cow buffalo should be checked for pregnancy?

(*a*) Between 45 to 60 days.

(*b*) Between 61 to 90 days.

(*c*) More than 90 days.

(*d*) Any other/No answer.

7. After how many hours of birth, a newly born calf should be fed colostrum?

(*a*) Within 2 hours.

(*b*) Between 3 and 4 hours.

(*c*) More than 4 hours or after dropping the placenta whichever is earlier.

(*d*) Any other/No answer.

8. What should be fed to a crossbred cow daily?

   (*a*) Green fodder + compound feed or concentrate mixture (prepared at home in proper proportion) + mineral mixture and feed supplement more than thrice a day.

   (*b*) Green fodder + concentrate mixture (prepared at home in proper proportion) + salt thrice a day.

   (*c*) Fodder and feed (whatever available) - *ad lib.*

   (*d*) Any other/No answer.

9. What are the main contagious diseases of cattle against which vaccination should be done?

10. What are the important symptoms of Foot and Mouth Disease (FMD)?

    (*a*) Initially there is high fever.

    (*b*) Sluggishness.

    (*c*) Reduced feed consumption.

    (*d*) Profuse salivation from the mouth.

    (*e*) Lameness

    (*f*) Smacking of the lips.

    (*g*) Abrupt reduction in milk yield.

11. What are the important symptoms of Haemorrhagic Septicaemia (HS)? Please tick the symptoms known to you.

    (*a*) There is short, rapid and painful respiration in initial stage.

    (*b*) Animal becomes restless, off-feed and start shivering.

    (*c*) Mostly swelling starts in head, throat, neck and chest region, out very extensive in throat region.

    (*d*) Body temperature rises sharply with rapid breathing.

(*e*) Watery discharge from the eyes and nose.

(*f*) The severe stage is marked by laboured breathing with peculiar grunting sound.

12. What are the important symptoms of mastitis disease? Please tick the symptoms known to you.

(*a*) Swelling of one or more quarters/teats.

(*b*) Pain in affected quarters/teats.

(*c*) Flakes or clots with the milk at initial stage.

(*d*) Pus, blood clots, turbid thin milk in advance cases.

13. What should be done to maintain the cleanliness of the pucca cattle shed/house?

(*a*) Sweeping and washing the floor daily in the morning, and white washing at shed at least twice a year.

(*b*) Sweeping and washing the floor once a day.

(*c*) Sweeping once in a day.

(*d*) Any other/No answer.

14. How much dry period one should allow for a lactating pregnant crossbred cow?

(*a*) 46–60 days before expected calving.

(*b*) 31–45 days before expected calving.

(*c*) 30 days or less before expected calving.

(*d*) Any other/No answer.

15. At what age one should get his crossbred calves dehorned?

(*a*) At the age of 7–15 days.

(*b*) At the age of 16–30 days.

(*c*) At the age of 31 days or more.

(*d*) Any other/No answer.

16. At what age one should get his male calf castrated?
    (*a*) Between 1 and 2 years of age.
    (*b*) More than two years but upto 3 years of age.
    (*c*) Above 3 years.
    (*d*) Any other/No answer.

H. ADOPTION OF DAIRY FARMING PRACTICES

Sr. No.	*Practices*	*Always (3)*	*Sometimes (2)*	*Never (1)*
Area - I : Breeding				
1.	Keeping a cow/buffalo			
2.	Keeping watch on oestrous cycle and heat symptoms of cow buffalo.			
3.	Getting covered the cow/buffalo within 60–90 days after calving.			
4.	A.I. at proper time of heat.			
Area - II : Feeding				
1.	Feeding colostrum to new born calf within 1-1½ hrs of birth.			
2.	Feeding colostrum continuously to new born calves upto 5 days of its birth.			
3.	Feeding concentrate mixture on the basis of milk production.			
4.	Providing green fodder to animals.			
5.	Feeding advance pregnant animals with extra 1-2 kg of concentrate over the above of maintenance ration.			
Area - III : Management				
1.	Practising loose housing system/stall feeding with grazing.			
2.	Providing clean and fresh water to animals.			
3.	Keeping the advanced pregnant animals separate from herd.			
4.	Keeping the calf and its mothers at warm place just after calving.			

(Contd...)

Sr. No.	*Practices*	*Always* (3)	*Sometimes* (2)	*Never* (1)
5.	Practising a dry period of 90 days (cow/ buffalo).			
6.	Maintaining cleanliness in animal shed.			
7.	Practising dehorning in calves at the age of 7–31 days.			
8.	Practising castration of male calves within 2 years of age.			
Area - IV : Health Care				
1.	Practising vaccination timely against the contagious disease (particularly - HS, FMD, BQ, Anthrax, etc.).			
2.	Segregating the diseased animals particularly in case of contagious disease.			
3.	Practising pesticides (Malthion/BHC, etc.) for the provision of ticks and kites, etc.			
4.	Practising deworming in calves for the prevention of internal parasites.			
Area - V				
1.	Are you adopting high yielding varieties of fodder crops?			
2.	Growing of Jowar/Bajra/Barley/Gram + Sun hemp for fodder crops.			
3.	Applying manures and fertilizers in fodder crops.			

I. KNOWLEDGE OF FARMERS ABOUT THE DEPARTMENT

1. What is the main aim of the department operating in your area?

    (*a*) To improve the indigenous breeds of cattle and buffalo.

    (*b*) To increase the milk yield.

    (*c*) To provide better veterinary aid.

    (*d*) All above.

2. Since how long the department is in operation in your area?

3. What facilities do the department extend to you?

   (*a*)

   (*b*)

   (*c*)

   (*d*)

   (*e*)

4. The department is meant to extend its service to

   (*a*) Large dairy owners.

   (*b*) Small dairy owners.

   (*c*) Marginal dairy owners.

   (*d*) All types of dairy owners.

5. Do the department mean to generate the employment opportunity to the educated unemployed youth? Yes/No

6. Do the department is concerned with fodder development in your area? Yes/No

7. Do you know that department organise camps (deworming camp, fertility camp, etc.)? Yes/No

8. Do you know, the department also conduct campaign related to improved dairy farming? Yes/No

9. Are you aware of the adaptive trials, demonstrations conducted by the department? Yes/No

ANNEXURE - VI

INDEX TO MEASURE FUNCTIONAL LINKAGE BETWEEN RESEARCH AND EXTENSION

Sr. No.	*Parameters*	*Always*	*Sometimes*	*Rare*	*Never*
1	*2*	*3*	*4*	*5*	*6*
A. Communication Linkage					
1.	Through on-station formal activities, *e.g.*, workshop, seminar, symposia, meetings, conferences.				
2.	Field activities, *i.e.*, field days, camp, campaign, demonstration, trial, surveys, etc.				
3.	Official correspondence.				
4.	Dairy mela, exhibition, livestock show, etc.				
5.	Farmers' training programme.				
6.	Publications.				
B. Collaborative Professional Activities					
1.	Performance of on-station and field activities, *i.e.*, seminar, workshop, mela, survey, field visit, etc. jointly.				
2.	Formulation and execution of field research/extension projects jointly.				
3.	Exchange of personnel and technical knowhow.				
4.	Identification of farmers' problems and evaluation of solution for the identified problems jointly.				
5.	Joint publications.				

1	2	3	4	5	6
C. Planning and Decision Making					
1.	Identifying the research problems and setting research priorities.				
2.	Deciding the priorities and objectives for extension activities/programmes.				
3.	Planning of the execution of research/extension activities.				
4.	Deciding the essential resources required for any/specific research/extension activity.				
5.	Planning and deciding the monitoring and evaluation of any research/extension activity.				
D. Implementation and Evaluation					
1.	Arranging the resources (materials/inputs/aids, etc.) for carrying out the research/extension activity.				
2.	Carrying out the planned programme/project phase-by-phase in a coordinated manner.				
3.	Monitoring and evaluation of the programme/project.				
4.	Modifying the programme based on concurrent as well as final evaluation of the programme/project.				
E. Training					
1.	Training need assessment of one-another.				
2.	Developing the training programme and course curriculum.				
3.	Participation in the actual training programme.				
4.	Monitoring and evaluation of training programme.				

(Contd...)

1	*2*	*3*	*4*	*5*	*6*
F. Supply and Services					
1.	Assessment of the requirements of technical inputs and services.				
2.	Technical verification of the inputs like semen, vaccines, medicine, fodder seeds/slip, etc.				
3.	Rendering technical services during field programme.				

ANNEXURE - VII

INDEX TO MEASURE FUNCTIONAL LINKAGE BETWEEN EXTENSION AND FARMERS

Sr. No.	*Practices*	*Always*	*Sometimes*	*Never*
A. Communication Linkage				
(*a*)	Information dissemination/seeking behaviour :			
1.	Visiting the farm/home/department.			
2.	Advisory letters.			
3.	Field days/demonstration/trial.			
4.	Training programme.			
5.	Meeting/group discussion, etc.			
6.	Leaflets, pamphlets, *dairy samachar*, newsletters.			
7.	Livestock show, dairy mela.			
8.	Camps, campaign and calf rallies.			
9.	Radio broadcast and telecast.			
(*b*)	Feedback :			
1.	Receiving/passing farm level problems from/to the farmers/department.			
2.	Getting/giving impression about the communication methods followed.			
3.	Getting/giving reaction during demonstration, trials, etc.			
4.	Getting/giving impression about the technical inputs and services.			
5.	Receiving/offering the opinion about the technical advices.			

(*Contd...*)

Sr. No.	*Practices*	*Always*	*Sometimes*	*Never*
6.	Receiving/giving opinion after the termination of camp, campaign, etc.			
7.	Getting/giving opinion about the training programme.			
8.	Receiving/writing the views about the literature.			
B.	**Planning and Decision Making**			
1.	Identification of farmers' problems and needs.			
2.	Setting the agenda and priority for field extension services.			
3.	Deciding the working/final plan and essential resources for any field extension activity.			
4.	Planning of the monitoring and evaluation of any extension activity.			
C.	**Implementation and Evaluation**			
1.	Arranging the site available for any field work.			
2.	Arranging the resources/materials (other than technical inputs)			
3.	Participate physically in the field work as camp, campaign, survey, trial, demonstration etc.			
4.	Mobilising the people to participate in the field programme.			
5.	Monitoring of the project/programme.			
6.	Evaluation of the project/programme.			
D.	**Supply and Services**			
1.	Assessing the requirements of technical inputs and services for field work.			
2.	Extending/availing the technical advices and services.			
3.	Supplying/receiving the quality inputs adequately, timely and at a reasonable price.			

(Contd...)

Sr. No.	Practices	Always	Sometimes	Never
4.	Cooperation during field supply of TIPs and services.			
E. Training				
1.	Training need assessment.			
2.	Developing the course curriculum.			
3.	Arranging the resource (physical, etc.) for the training programme especially the off campus training.			
4.	Physical participation in actual training.			
5.	Evaluation of the training programme.			

ANNEXURE - VIII

INDEX TO MEASURE FUNCTIONAL LINKAGE BETWEEN RESEARCH AND FARMERS

Sr. No.	*Parameters*	*Always*	*Sometimes*	*Rare*	*Never*
1.	Through personal contact, *i.e.*, visits, advisory letters, office call, etc.				
2.	Through meeting, group discussion, training, etc.				
3.	During campaign, camp, trial, demonstration, etc.				
4.	During field days and field visits.				
5.	During mela, exhibition, calf rallies, livestock show, etc.				
6.	Through publications (leaflets, pamphlets, handouts, etc.).				
7.	Through broadcast and Telecast.				

ANNEXURE - IX

CONSTRAINTS EXPERIENCED BY THE RESEARCH PERSONNEL, EXTENSION PERSONNEL AND DAIRY FARMERS IN MAINTAINING LINKAGES AMONG THEM

A. CONSTRAINTS IN RESEARCH-EXTENSION LINKAGE

Kindly enumerate the problems/reasons in maintaining linkages/non linkages with research/extension personnel. Some of the areas of constraints have been identified. Please write against each head :

(*i*) Organizational constraints :

(*a*)

(*b*)

(*c*)

(*ii*) Budgetary constraints :

(*a*)

(*b*)

(*c*)

(*iii*) Psychological/motivational constraints :

(*a*)

(*b*)

(*c*)

(*iv*) Communication constraints :

(*a*)

(*b*)

(*c*)

(*v*) Technological constraints :

(*a*)

(*b*)

(*c*)

B. CONSTRAINTS IN EXTENSION-FARMERS LINKAGE

B.1 Constraints Experienced by the Extension Personnel

Kindly enumerate the bottlenecks experienced by you in maintaining linkage with the dairy farmers. Please write the same against each heads :

(*i*) Organizational constraints :

(*a*)

(*b*)

(*c*)

(*ii*) Budgetary constraints :

(*a*)

(*b*)

(*c*)

(*iii*) Mobility/conveyance constraints :

(*a*)

(*b*)

(*c*)

(*iv*) Motivational/psychological constraints :

(*a*)

(*b*)

(*c*)

(*v*) Technical constraints :

(*a*)

(*b*)

(*c*)

(*vi*) Farmers' related constraints :

(*a*)

(*b*)

(*c*)

B.2 Constraints Experienced by the Dairy Farmers

(*a*)

(*b*)

(*c*)

(C) CONSTRAINTS IN RESEARCH-FARMERS LINKAGE

C.1 Experienced by the Research Personnel

(*i*)

(*ii*)

(*iii*)

(*iv*)

C.2 As Experienced by the Dairy Farmers

(*i*)

(*ii*)

(*iii*)

(*iv*)